Opec and the international oil industry

A changing structure

for
MY DAUGHTER
DOUNIA

OPEC AND THE INTERNATIONAL OIL INDUSTRY

A changing structure

BY

Fadhil J. Al-Chalabi

Published by Oxford University Press
on behalf of the Organization of
Arab Petroleum Exporting Countries (OAPEC)
1980

Oxford University Press, Walton Street, Oxford OX2 6DP

OXFORD LONDON GLASGOW
NEW YORK TORONTO MELBOURNE WELLINGTON
KUALA LUMPUR SINGAPORE JAKARTA HONG KONG TOKYO
DELHI BOMBAY CALCUTTA MADRAS KARACHI
NAIROBI DAR ES SALAAM CAPE TOWN

British Library Cataloguing in Publication Data

Al-Chalabi, Fadhil J
OPEC and the international oil industry.
1. Petroleum industry and trade
I. Title II. Organization of Arab Petroleum Exporting Countries
338.2'7'282 HD9560

ISBN 0–19–877155–X Pbk.

ISBN 0–19–877167–3

Phototypeset in V.I.P. Palatino by Western Printing Services Ltd, Bristol and Printed in Great Britain by Billing & Sons, Ltd, Guildford, London & Worcester

Foreword

The origin of this significant and timely book was a lecture on the structural changes in the international oil industry prepared by Dr. Al-Chalabi for the OAPEC annual course on the *Fundamentals of Oil and Gas*. It was given in Arabic in February 1978 when the author was Assistant Secretary General of OAPEC. The author has succeeded in expanding and updating the basic aspects of the subject. He has managed, in this relatively short book, to survey and analyse the main structural changes in the international oil industry during the last 35 years, with special emphasis on the revolutionary changes of the 1970s.

The book clearly demonstrates the significance of these changes in terms of their impact on the power of control and decisions concerning the ownership of oil resources, their marketing outlets and price policy. It explains how the concession system and the various consortia arrangements among the transnational oil companies gave them maximum freedom and flexibility in the control and management of crude oil production and pricing. Their vertical and horizontal integration enabled them to control and manage upstream and downstream operations on a global basis and with the effectiveness of a powerful international Cartel. The major oil companies as a group controlled the supply of crude oil, its transportation, its processing plants and its marketing and distribution system. It was a closed circuit where the companies made all the major decisions for producing and consuming countries. There is now general consensus, even within the oil companies, that they generally made their decisions in favour of the consuming rather than the producing countries. The result was a huge transfer of cheap oil resources from the poor producing countries to the relatively affluent oil importing industrial countries.

It was against this background that OPEC was established in 1960. The entrance of independent and national oil companies from the consuming countries during the sixties provided OPEC with the opportunity to manifest its presence with increasing effectiveness. It also marked the beginning of major structural changes in the oil industry which paved the way for the revolutionary developments of this decade.

The reader will find an interesting and highly informative explanation of the causes and implications of these and other changes. He will also find candid recognition of the structural problems and policy risks associated with the fragmentation of the upstream operations among sovereign states. The need for more cooperation and coordination among the oil exporting countries is clearly demonstrated.

It is fortunate that this kind of book is written by a man of Dr Al-Chalabi's academic qualifications and experience. It is rare that executives of his position find the time or the freedom to write on the critical aspects of the international oil industry. Those who write books on the oil industry are usually either of academic background, or journalists, politicians and former oil executives who generally place heavy emphasis on their particular personal experience in their company and in a particular country. It is also rare for readers from the oil importing countries to find a book in their own language authored by an executive from the oil exporting countries. Dr Al-Chalabi practised his profession at the technical and policy levels in his country and in the context of regional and international organizations. His vast experience with the work of OPEC and OAPEC and his personal association with the leading personalities and decisions of these organizations has enabled him to examine OPEC and the oil industry with a combination of professional competence and practical knowledge.

I have greatly enjoyed reading this book and I trust that others will feel the same way about it. Regardless of whether they agree or disagree with its main conclusions, I feel sure that they will find the main aspects of the international oil industry revealed in brief, simple and direct style.

ALI A. ATTIGA
Secretary General of OAPEC

Preface

The original project for this book was to publish a translated version of my lecture on the 'Major structural changes in the international oil industry', published in Arabic in the form of a booklet during the summer of 1978. However, the work looked too succinct and sketchy to give to readers in the consumer countries an adequate understanding of a producer's point of view on the great developments in the international oil industry and the role of OPEC. On the other hand writing afresh a fully-fledged book on the subject seemed to me unfeasible in practice, due to the many work committments I had to meet. I therefore took a mid-way solution: to undertake only such work on the original booklet as was necessary to explain, expand, update, correct and write new sections or parts thereof. This is why many important additions were made in the form of footnotes rather than insertions in the text. The result is this book as it stands and I hope that it will meet the minimum requirements of comprehensiveness and clarity. But I am the first to recognize its imperfections. I should note here that the book was largely written during 1979 and any mention of 'the current decade' refers, therefore, to the 1970s.

The major part of the credit for this book should go to the Organization of Arab Petroleum Exporting Countries especially the Secretary General, Dr. Ali Ahmad Attiga, whose ardour, imaginative capacities and unshakeable belief in co-operation among the oil producing countries have played a tremendous role in strengthening that pioneer Arab organization. The idea that I lecture on the structural changes of the industry was his, and the encouragement and support he gave me were extremely helpful. Also the remarks he made on the text helped me in clarifying some of the issues. I must therefore record here how indebted and grateful I feel towards him. Dr. Walid Khadduri, Director of the Information Department of OAPEC, should also receive part of the credit for this book. Without his dynamism and persistent following up, the book could easily have been delayed amidst my very heavy work load.

I must also record here my deep thanks to the OPEC Secretariat which greatly helped in providing the necessary data, figures, tables, etc., which enabled me to update and expand the book, besides supplying the secretarial work, including some editing, stenography, typing, etc.

A word of thanks and appreciation go to my friends and colleagues, especially Dr. Abdul Amir Al-Ambari, President of the Iraqi Fund for External Development, Mr. Adnan Al-Janabi, Head of the Economics and Finance Department of OPEC, and Mr. Sabri Kadhim, Head of the Energy Studies Department of OPEC, who all, in one way or another, have gone through the text, whether the original Arabic version or part of the expanded English one, and from whose remarks and suggestions I greatly benefited.

It goes without saying that all the views expressed in this book are personal and strictly my own and I am entirely responsible for them. They, therefore, do not reflect those of OPEC, in which I serve or OAPEC, in which I used to serve or any of their organs; neither do they, of course, reflect any of the OPEC member countries' views.

FADHIL AL-CHALABI
Vienna
November/December 1979

Contents

Introduction

No industry in the world has witnessed as many radical changes as those that have occurred in the international oil industry, particularly those affecting the extraction of crude oil and its exportation from the principal producing/exporting regions. Rapid changes in recent years in the relation between the oil producing countries and the large energy establishments of the oil consuming/importing countries, especially the multi-national oil companies, have changed the structure of the industry, and have caused world oil relationships to enter a new phase which has significant consequences for the international economic order.

A chain of structural changes has radically altered many areas and aspects of the oil industry, in particular those pertaining to the patterns of oil extraction in producing/exporting areas and the system of international marketing of oil, as well as the pricing of oil, which is the most effective and important channel for wealth transfer and sharing between raw material exporters and importers.

However, the pivotal point for all these changes was the growing and forceful exercise by the oil exporting countries of the concept of the inalienable right of a state to permanent sovereignty over its natural resources. The effective practice of this right was the real impetus for all the recent radical developments that have occurred in the international oil industry.

The facets of these changes are numerous and interrelated, but their developments can be summarized under four headings.

(1) *The system for the extraction and development of crude oil* The system of oil concessions inherited from the old colonial rule ultimately collapsed through the replacement of the multi-national major oil companies in the ownership and management of the upstream extraction phase of the oil industry by the governments of the producing countries.

(2) *The system for marketing crude oil*
Major developments occurred in the structure of the world market for crude oil resulting in the relative eclipse of the major oil companies in the sphere of international marketing of crude oil in favour of the national oil companies of producing countries.

(3) *The system for oil pricing*
This comprises all developments in the sharing of the accumulated economic wealth generated from oil resources, and the determination of the oil producing countries' share in that accumulation through the pricing of oil. Successive changes in the pricing system have finally led to the replacement of the oil companies by the oil producing countries in determining oil prices.

(4) *Basic issues of OPEC price administration*
With OPEC assuming the responsibility of solely administering the price of oil exports from the main producing areas, a new set of price relationships has emerged. The problem of oil prices has changed from a government-company relationship to an international and also intra-OPEC relationship.

Fundamental characteristics of the international oil industry

If the series of structural changes in the oil industry have had an importance and a radical nature not previously experienced by any other industry, then the roots of this transformation must lie in the very nature and characteristics of the oil industry itself.

One of the basic characteristics of the industry, and one which has played, and will continue to play, a decisive role in its structural changes, is the fact that oil is an exhaustible natural resource, in the sense that the continued extraction of oil from a reservoir will lead eventually to its total depletion. In other words, an extracted barrel cannot be replaced by another barrel, except by an exploration effort which might or might not end in the discovery of new reservoirs, or by enhancing efficiency in the recovery of oil already in place—and actually adding new reserves. This means that, if adequate continuation of reserves is desired, the barrel extracted today demands the further expenditure of large sums of money, because of the great capital risk involved in replacement; a risk which increases with the passage of time, since the more oil is extracted the less likely is it that new oil will be found.

This feature has a special significance for the producing/exporting regions of the Middle East and Africa, the most important of these being the member nations of the Organization of the Petroleum Exporting Countries, since the reserves-to-production ratio (the measure for the lifespan of oil reserves), in the Middle East in particular, has fallen from the highest level of the fifties (130 years at the 1957 production rates) to the lowest level of the seventies (38 years at the 1977 production rates), despite the fact that the size of the proven recoverable reserves in that area has multiplied approximately more than three times during that same period.

A second feature of the industry is that it is mostly oriented to world markets, and that its growth depends on international trade. If we leave out of account the Soviet Union, and to a much lesser extent the United States of America which has become heavily dependent on oil imports, the oil industry in the rest of the world relies basically on the exportation of oil from areas where oil consumption is low because of low levels of economic and social development, to areas with no significant domestic oil production but with high levels of industrial development and consequently high rates of oil consumption.[1]

The current world trade in oil accounts in value for about one fifth of total international trade. However, the bulk of that trade (85 per cent) takes the form of crude oil, as

1. The Soviet Union and the United States are exceptional for, while they both enjoy a high level of internal consumption, their great oil producing potential has nevertheless isolated them for a long time from the international oil trade. The Soviet Union was, and still is, a net exporter of crude to the world outside the Socialist Bloc (in the order of 1·3 million barrels per day). In the US however, the oil situation has changed drastically in the last eight years as current imports to this highly energy-intensive country account for about half of its consumption. (The US currently imports more than 8 million barrels per day). Until the turn of the last decade the bulk of US oil consumption was met by internal production, which was protected against heavy imports of cheap foreign oil by various means of quantitative controls.

consuming importing countries refine at home imported crude oil. This feature represents a drastic reversal in the pattern of world oil trade during the pre-war period when more than two-thirds of oil exports took the form of refined products exported from 'source' refineries in the major producing areas in the Middle East and the Caribbean. The post-war period witnessed drastic changes in the locational distribution of world refining capacity. Whereas before the war Western Europe refined at home only one-third of its total oil consumption, its current refining capacity exceeds by far its oil consumption.

Another characteristic of this industry is its steady and sustained growth, resulting from the fundamental changes which have occurred in the economies of the industrialized countries, particularly since the Second World War, and the transformation of their energy structure from a basic reliance on coal to a basic reliance on oil and gas. This transformation was due to the properties of oil, which made it the fuel most suitable for the technological innovations and changes necessary for the Western economy. Oil—a scarce commodity increasingly used as an industrial input and feedstock for the chemical and petrochemical industries[2]—continues to be used mainly as the primary fuel for industry, for transportation, for domestic purposes and services. This is why the rate of growth in oil consumption in Western Europe and Japan has been high, to the point of doubling consumption every five to seven years in the post-war period, at a time when world oil consumption was doubling on average every ten years.[3]

This turns oil into a strategic commodity closely linked to

2. The consumption of oil as an industrial feedstock (raw material input) for those industries accounts for less than 15 per cent of total world oil consumption.
3. This pattern of energy growth in the developed countries has changed drastically since the oil revolution of 1973–4. In 1978 oil consumption in Western Europe was less in absolute terms than in 1973, i.e. an average annual decrease of about 1 per cent. The reversal was greater as far as their net oil imports were concerned since the development of North Sea oil resulted in the imports from the OPEC area dropping further in absolute terms. The situation in the US, on the other hand, was different: consumption continued rising albeit at lower than historical rates. Between 1973 and 1978 the annual average growth rate was 1.7 per cent, against a historical rate of over 4 per cent.

world power politics because the security of its supply and the level of its price plays a crucial role in the process of economic growth, and because the technological changes and the tremendous increase in industrial production has brought the world economy, especially in the highly industrialized countries, to a point of total dependence on oil. Economic growth, and consequently increased social welfare on which these countries' economic and political systems depend, can no longer be achieved without oil.

Another feature which indicates the uniqueness of the oil industry is its integrated nature, for oil produced cannot be made available to end consumers without being passed through various phases, each of which generates new added values. There exists an organic link between each phase, with all phases being interdependent. The extraction of crude oil cannot be begun before the process of exploration for and discovery of oil has been undertaken. The crude cannot serve an end or be used by the consumers without being transported to the areas of consumption (whether by pipeline or tanker), where it is passed through various refining operations and offered to consumers in the form of refined products. In consequence, an investment in any one phase of the industry depends largely on the nature and size of investments in the other phases; the amount of investment at each phase forms the market for the preceding phase and so on.

Another important characteristic of the oil industry is that it is highly capital- and technology-intensive. The capital costs (expenditures necessary for installing production capacities), particularly at the phase of crude discovery and development, constitute a very high proportion of the total cost involved in the oil industry in general. With this high ratio of capital expenditure, operating costs of the industry (to meet expenditures for the current requirements of production) are regarded as being very low in relation to the overall cost structure. The high capital risk at the phase of initial investment in oil exploration involves huge outlays in searching for oil, especially in drilling exploratory wells, which may turn out to be dry holes. Less risky, however, are the investments involved in developing oilfields, including transportation, de-gasing stations, export facilities, the various processing facilities, etc.

As the history of the international petroleum industry shows, this particular feature played an important role in anchoring the structure of the industry to an oligopolistic basis (operators being few in number) because short-term market factors led, in the absence of price agreement among operators, to competition that would be highly damaging. An example is the price wars between the companies in the twenties. In order to preserve his market share, an operator might undertake massive price cuts that would bring his price so near those low levels of operating costs, that the cuts would seriously damage the growth of the industry itself.

PART I

The system for crude oil extraction: the fundamental structural changes in the ownership and management of the 'upstream' phase of the oil industry in the producing countries

From the start of oil exploitation in the principal producing/exporting areas of the world at the beginning of this century, the predominant system for investment in oil was characterized by two basic features.

First, the system for petroleum concessions, which were held by foreign companies and covered more or less the entire territories in the producing areas usurped the right of the state in those countries to permanent sovereignty over its petroleum resources. Because the foreign concession-holding companies controlled the oil operations it was, in practice they and not the state of the host countries, who exercised that sovereignty.

Second, the inter-company arrangements among the Seven Sisters (the International Petroleum Cartel) and their joint and collective assumption of concession rights enabled them to exert total control over the process of oil extraction in the upstream.

The oil concession system was the fruit of the old colonial rule, under which the Great Powers vied with one another in carving out zones of influence in the world. This was a process which lasted from the beginning of this century till the end of the First World War and which drew to a close with the victorious Powers confirming their control over the regions in the Middle East that lay under colonial influence. It was the British and French who were first to carve up those areas in accordance with the San Remo and other treaties, and only subsequently did American influence emerge via what was called the Open Door Policy.[4]

The concession system involved the exclusive right of the concessionaire to exploration, extraction, and exportation of oil, so that no other investor could compete within the area of concession. In its early days the system concerned only specific and defined lots of land, and the concession-holder was liable to undertake a minimum expenditure for oil search and exploitation within these areas and within a definite time period, after which he had to relinquish the lands to the state in the event of non-discovery of oil or the non-development of discovered oilfields. The agreement with the Turkish Petroleum Company in 1925 was an expression of this so-called principle of relinquishment. In

4. In his book *The Middle East Oil and the Great Powers* (Transaction Books, 1973), Benjamin Schwadran gives an extremely interesting account of inter-European conflicts on the Middle East oil, especially Iraq. The creation of the Turkish Petroleum Company to exploit Iraqi oil resources was the first form of the consortia of multi-national shareholders representing the post-war political compromises. After the French took over the German share promised to the Deutsche Bank by the Ottoman Empire before the war, a *modus vivendi* was reached with the British represented by Anglo-Iranian and the Royal Shell. A new political status quo had to be created, following the increasing pressure of the United States to secure a share in the Middle East oil. The Americans being among the victorious Powers of the war, claimed that the door to the Middle East oil should be open to all the partners of the victory. Thus the amendment of the shares in the Turkish Petroleum Company, which was given the concession right of exploiting Iraqi oil in the territories east of Tigris and which later became Iraq Petroleum Company. Accordingly the United States was given 23.75 per cent of the shares, to be equally divided by the two American multi-national oil companies, The Standard Oil of New Jersey (later to become Esso and then Exxon) and Mobil. The French had, through their government-backed Compagnie Française de Pétroles, another 23.75 per cent of the shares, while Anglo-Iranian (later to become British Petroleum) held a similar share as well as the Royal Dutch/ Shell, thus leaving the remaining 5 per cent to Gulbenkian. This pattern of sharing had to be applied elsewhere in the area, but only within the borders of the Redline Agreement, so that none of the shareholders could enter individually. In the areas outside the Redline Agreement, none of the companies was under any obligation to develop oil resources jointly with the others. So the American companies succeeded in securing the exclusive rights of exploiting Saudi Arabian oil resources through a consortium later created and given the name of Aramco, and with Exxon, Texaco, Socal having each 30 per cent and Mobil the remaining 10 per cent shares. Helmut Mejcher's fascinating analysis, in his book *Imperial Quest for Oil: Iraq 1910–1928* (Ithaca Press, 1976), of the conflicts among the Great Powers to dominate Iraqi oil, is based on more recent research and documentation.

practice, however, the system of concession subsequently evolved to cover all the territories of the nation in which the companies operated.[5]

The concession system made the holder the sole arbiter of the volume and nature of investment in the host country, the choice of areas for exploitation, the determination of exploration plans, the development of oilfields, the production levels, the size of the necessary production facilities, exportation and transportation capacities, etc. In practical terms this deprived the state of the right to interfere in any of these vital matters and limited its role merely to that of collecting taxes, so that the legal relationship between the concessionary company and the state was purely fiscal. In accordance with the general pattern of the concession agreements, so-called commercial expediency was the only criteria observed by the concessionaire when taking investment and production decisions. Obviously such criteria would secure for the concessionaire the highest profits without entailing any liability to meet requirements for the economic and social development of the host country. Nor did they entail any obligation towards the development of the indigenous oil sector itself.

This meant that the considerations which dictated the size of investment and investment plans in the concession areas were ultimately linked to the needs of economic

5. The principle of relinquishment was admitted in the agreement of 1925 with the Turkish Petroleum Company Ltd, whereby the concessionaire held 24 plots which had to be returned to the government after a certain period, in case of failure (the plot system). By the conclusion of the 1931 agreement with the IPC (which replaced the TPC), this principle was abandoned and the concessionaire was given the entire East of Tigris northern territory of Iraq without any obligation ever to invest in exploration or development. Following the conclusion of the agreement with the Mosul Petroleum Company Ltd which was owned by the same shareholders of IPC (BP, Shell, CFP, Esso, Mobil, and Gulbenkian) the entire West of Tigris northern territory of Iraq was given up under a separate concession. By 1938, when a third agreement was signed with the Basrah Petroleum Company Ltd. (a new company owned by the same shareholders) covering the entire southern territory, the whole of Iraq territory was exclusively covered by the Cartel. Details about the consortium of the Iraq Petroleum Company exploiting Iraqi oil are found in Mejcher (op. cit.). Also S. H. Longrigg in *Oil in the Middle East: Its Discovery and Developments* (Oxford University Press, 1967), and Schwadran (op. cit.), give useful details about the story of the Iraqi oil, the first that set a pattern in the Middle East.

growth in the industrialized world as reflected in world oil requirements. In other words, the concession system by its very nature supported the dualistic structure of the producing economy, which meant that the oil sector in the host country was cut off from the development of the national economy and rather closely linked to the world economy. We therefore find that the high degree of technological progress so visible in this sector did not extend to the other sectors of the national economy, which remained in a state of economic stagnation for a long period.

The other most significant aspect of the international oil industry at the extraction stage (the upstream) was the predominance of the inter-company oligopolistic arrangements (the International Petroleum Cartel) which secured the joint entry of the major oil companies into the oil producing regions for the joint development of oil resources. This was made possible only through the creation of consortia whose shareholders were the companies themselves. It was these consortia, and not their shareholders, that held the concessions in virtually all the producing areas of the Middle East. The predominance of this system had great influence on the growth of the international oil industry, on the relationships of the political centres of world power and on the determination of the oil producing countries' status, especially pertaining to their share in the increased wealth created by this exhaustible commodity, oil.

It is not our intention here to run through the history of the formation of the Cartel, and of the secret agreements between the companies involved, most notorious of which were the Redline Agreement and the Achnacarry Agreement. But one cannot avoid mentioning the distinctive features of this monopoly and its effects on crude oil exploitation operations in the producing countries. While the Cartel companies had a joint stranglehold on the production of crude, they individually had control over downstream operations, such as the transport of crude oil, its refining, and the networks for distribution to the ultimate consumer throughout the world outside the centrally planned economies (and also, though to a lesser extent, the United States where the control exercised by these companies was less than in other regions). In other words, these companies, individually and collectively, were vertically integrated, in the sense of having access to investment and ownership at all stages of the oil industry, starting from crude oil production to distribution of the refined products to the final consumer. Their investment decisions at any phase of this industry were closely bound up with all the other interdependent phases.

However, the principal feature of the international Cartel in the producing regions was that the companies entered collectively into the crude oil production operations, as opposed to downstream operations in which each company individually pursued its investments independently from the others. The companies made their collective entry into the crude oil production phase by means of an investment coalition, a strictly operational and 'non-profit making' consortium whose task was confined to making the crude oil available for lifting by the shareholding companies at cost plus a nominal amount to cover administrative expenses (the famous one shilling per ton plus cost). The significance of such an arrangement was that the concession in each production area was legally held by the consortium and not its shareholders, so that the legal relationship between host government and companies was exclusively confined to jointly-owned operational entities, and not their owners, the multi-nationals.

The consortium phenomenon, which was widespread in almost all the producing regions (except for some recently developed areas such as Libya and Nigeria), caused the interests of the eight major companies therein to be interlinked at the crude oil production phase because each company had some share in one or more of the concession-holding consortia in virtually all the major oil producing countries, especially in the Middle East.[6] In this way the shareholding companies were effectively horizontally inte-

6. The Cartel was composed of the British Petroleum (with majority share of the British government), the Royal Dutch/Shell (an Anglo-Dutch joint venture) and five American major oil companies, Exxon (previously called Esso), Texaco, Gulf Corporation, Socal and Mobil. The French company, Compagnie Française de Pétroles, occupied a minor *strapontin* seat in the oil 'club'.

grated at this upstream phase of the international oil industry, as far as the oil producing/exporting countries were concerned.

Effects of the monopoly on the industry

The combination of the inter-company oligopoly arrangement and the concession system had great and far-reaching consequences on the oil industry, which can be summarized as follows:

> (a) Investment operations in the upstream, including exploration, the development of discovered oilfields, the construction of production and exporting facilities, etc., were very largely connected with downstream investment. This meant that the size and capacities of the global distribution networks for the final consumers on a world scale were actually what decided the volume and capacity of crude production in the upstream. Investments in the upstream operations formed part of the companies' worldwide integrated operations, and upstream operational costs, including the taxes and royalties paid to the host governments, were regarded as part of the overall costs of the industry's integrated system. Here the difference between the total proceeds from the sales of refined products all over the world and the total costs of all these worldwide operations, represented the true profit of the major oil companies. In other words, the companies' profits from their upstream operations only existed for bookkeeping purposes, mainly as a basis for taxation.

Those companies' percentage shares in the oil consortia in the Middle East were the following (in January 1972):

	Iraq	Iran	Saudi Arabia	Kuwait	Qatar	UAE (ADPC)
BP	23.75	40	–	50	23.75	23.75
Royal Dutch/ Shell	23.75	14	–	–	23.75	23.75
Exxon	11.875	7	30	–	11.875	11.875
Mobil	11.875	7	10	–	11.875	11.875
Gulf Corp.	–	7	–	50	–	–
Texaco	–	7	30	–	–	–
Socal	–	7	30	–	–	–
CFP	23.75	6	–	–	23.75	23.75

(b) The presence of oil consortia in all the producing areas and the fact that they were owned by the same shareholding companies provided them with the tools to disregard in practice the territorial sovereignty of the countries in which they were operating as far as oil operations and investment-making decisions were concerned. Let us take Esso (currently Exxon), as an example. It owned various percentage shares alongside the other multi-national oil companies (through operating consortia) in Saudi Arabia, Iran, Iraq, Venezuela, Qatar, Abu Dhabi, etc., at the same time as running its own refineries and distribution networks in Europe, the Far East, and the Western Hemisphere. The investment decisions made by Esso in any one producing country depended, therefore, not on the needs of that country but on the requirements of the company itself and its operations in all countries and at all phases of its oil operations. This meant, for instance, that what Esso decided on as regards production in Saudi Arabia depended on the one hand on the company's production potential in other producing areas and, on the other hand, on its downstream operations' crude oil requirements, as defined by the refining capacity owned and operated world-wide by Esso itself (through its various subsidiaries and affiliates). Thus the nature of the companies' operations was 'inter-territorial', as dictated by their individual and collective linkages in the various territories in which they were operating through the concession system. It was therefore 'inter-territoriality' that governed inter-company investment behaviour, while host countries were left to act merely as tax-collectors in their individual territories.

(c) It was a feature of the collective entry by the companies at the upstream phase in the producing areas that investment risks were shared among the companies. An investment failure in one region would be more than compensated for by successes gained in others. This had great importance in minimizing investment risks and in reducing costs, thereby helping the companies to increase their world-wide profits.

(d) The concession system and the presence of the Cartel at the upstream phase provided the companies with the tools for the planning and programming of crude oil production in the areas covered by the concessions. The primary aim was naturally to avoid any surpluses or shortages in their overall world-wide balance of crude oil supply and demand. No less important an aim of such production programming was to preserve the relative shares of the parties in the various consortia in an expanding market, with the ultimate intention of maintaining the status quo of each of these companies, (the famous 'As Is' rule of the Achnacarry Agreement). This joint production planning was restrictive in that it pre-empted any growth in the share of the smaller shareholding companies at the expense of the larger ones. There were several formulae for implementing such planning, the best known being the cyclical five-year production plans initiated by the Iraq Petroleum Company. This company prepared five-year investment and production plans, which were conceived five years in advance and were subject to annual revision. Under this plan the shareholders in the IPC consortium determined their future needs for the crude produced in the area covered by the concession on the strength of the forecast for the shareholders' requirements for crude as determined by the requirements of their refining and distribution capacities outside Iraq. They would 'nominate' in advance the quantities of crude oil that they would need in years to come. By totalling these quantities and adjusting them by applying certain restrictive rules (such as the five-sevenths rule), the final result formed the goals for crude oil production in Iraq, and hence the investment requirements and operations through which the five-year plans were implemented.[7]

Through such tools of production programming, the shareholding companies, singly and collectively, exercised complete control over the balance of their needs for crude against their own downstream capacities in refining and distribution; this meant that they had complete control over the crude oil supply for international trading as well as direct influence on world demand.

In spite of these arrangements, inter-company crude

7. The five-year plans pursued by the Iraq Petroleum Company in the planning of oil production in Iraq show the restrictive nature of the inter-company crude production arrangements, which actually prevented any further expansion of Iraqi crude oil production. The aim of applying some of the restricting rules, such as the five-sevenths rule, to Iraqi oil production was to prevent smaller companies, such as the Compagnie Française des Pétroles, which did not own large amounts of crude oil outside Iraq, from expanding into the production of cheap oil by way of increasing Iraqi production. This would have threatened the strong central position occupied by the stronger companies such as British Petroleum and Esso, which owned large shares of

imbalances existed within the Cartel in the sense that some of the partners owned more crude oil than they needed to satisfy their downstream requirements, and were called the crude 'surplus' companies, compared to other partners who owned less crude oil than their own downstream requirements, who were called 'deficit' companies. These imbalances were, however, handled through long-term lifting agreements among the partners so that the surplus companies 'sold' crude oil to the deficit companies, thus avoiding an overall imbalance in the companies' supply and demand.

This aspect of production planning played a major role in influencing the growth of the petroleum industry and the economic return of oil investments at the upstream phase of the oil operations in the producing countries, especially from the point of view of eliminating waste and losses, inter-company competition, partial utilization of plant capacity, etc.[8] On the other hand, these kinds of inter-company arrangements made great inroads into the producing countries' sovereignty, since the companies' investment and production decisions were made on optimized regional, and not strictly national considerations thus, in practice, by-passing the borders of national sovereignty. By resorting strictly to regional optimization of oil investments, company decisions were in many cases in direct conflict with the national needs of producing countries.

(e) The concession system did not imply a particular obligation on the concessionaire to invest in oil exploration or production, neither did it specify the minimum expenditure for development of the producing countries. The com-

crude oil production in other regions. It was in the interests of the larger companies that Iraqi production should not expand, a fact which explains the long stagnation of the oil industry in Iraq. The US Federal Trade Commission report *The International Petroleum Cartel, 1952* contains details of inter-company relationships within the Cartel concerning the amounts of crude oil available among the shareholders and the programming of crude production in the conventional OPEC areas. Likewise Professor Edith Penrose gives in her book *The Large International Firms in Developing Countries* (Allen and Unwin, 1968), Chapter V, an extensive analysis of the influence of integration and oligopolistic control over production of crude oil.

8. The high return on the investments of the companies in the producing countries of the Middle East compared with a low or negative return on some of their investments at the crude production phase in other, non-OPEC regions. The result was that the unusually low cost of the companies' investments in the OPEC areas formed the main source of finance for their investments outside these areas.

panies were therefore pursuing investment policies that allowed only such expenditures in any particular country as were necessary for their overall world-wide investment programmes, irrespective of the needs of that country for economic and social development. What occurred in Iraq provides a good example: the companies, having discovered the giant Kirkuk oil-field in the north and the Rumailah field in the south, did not make any further investments to discover or develop new fields. This policy, damaging to the national economy, can be explained by the hardline position taken by the Iraqi government which led to a continued deterioration in the companies-government relationship over what was in effect an expropriation of Iraqi rights. As a result the expenditure on oil search in Iraq was among the lowest in the world. It was obvious from this investment behaviour that the companies were intending to put economic pressure on Iraq.

Erosion of the oligopoly's control of the industry

This system of complete domination by the Cartel of upstream investment and extraction operations lasted for a long time and effectively remained in use until recently, despite the partial shake-up in the wake of the Suez Crisis (the concession system remained fully operational until the early seventies). The developments which led eventually to the total collapse of the whole system, involved the radical changes of the past seven years, which brought about a complete reversal in oil investment relationships and eventually the replacement of the foreign companies in the ownership and management of the oil industry by the governments of the producing countries.

The attempt at nationalization made by the Mossadegh government in Iran in the early fifties was the first world event, which, despite its failure, had a far-reaching political impact on the system and which led to its first shake-up. This event laid the real foundations for subsequent political developments in the region, and led to a widespread dissemination of the concepts of full control by the state over its oil operations, by government takeover (nationalization by legislation) or by mutual agreement with the companies. (Both concepts underlie the present shape of the industry).

Yet it was the subsequent developments, particularly during the sixties, that were effective in undermining the companies' domination, and enabled the oil producing countries gradually and effectively to extend their authority in the control of the oil industry in their territories. Perhaps the most significant development which contributed to this shake-up, and which played a great role later on in radically changing the structure of the industry, was the growing trend in the producing countries towards the establishment of national oil companies. These were designed from the beginning to be the instruments through which the state could exercise its rights over national resources. National oil companies of the producing countries were given the opportunity to invest (mostly in joint ventures with independent foreign oil companies) in certain areas, albeit marginal, which were either not covered by the concession (very small areas) or were voluntarily relinquished by the major companies by agreement with the governments of the producing countries as, for example, in Iran.[9] Later there were some more advanced forms of direct involvement by the national oil companies, especially in lands which were restored to the governments through legislation as, for example, in Iraq, when the government promulgated Law No. 80 of 1961. This kind of enforced relinquishment by legislation enabled the government to recover all lands covered by concessions, but not exploited or developed (leaving to the companies only such lands where they had effective operations). In Iraq all recovered lands were given

9. Following the CIA-backed coup which aborted the nationalization act and restored the companies' control over Iranian oil in the form of an international consortium (no longer exclusively dominated by the British but including a predominant American presence) some small areas previously covered by the old Anglo-Iranian concession were relinquished to the state by agreement with the consortium. These lands were given to the National Iranian Oil Company (NIOC), established by the Mossadegh government but not dissolved under the new regime. The new government did not formally abrogate or repeal Mossadegh's nationalization act but gave the newly-established consortium the right to exploit and operate in Iran's nationalized oilfields on behalf of NIOC, which remained the legal owner of the fields. The terms of the consortium's operations in Iran were such as to produce the financial and general results prevailing in the area. The relinquished marginal acreage in Iran was subsequently exploited by NIOC through independent foreign oil companies such as Phillips, Pan Am, ERAP, etc.

to the Iraqi National Oil Company (INOC), which started exploiting oil resources either jointly with some independent foreign oil companies or, later, directly with the technical assistance of the USSR.[10]

These national oil companies would not in fact have been able to play their historic role in oil exploitation in the producing countries without other important developments on the international scene that effectively helped to shake the Cartel's hegemony over the international petroleum industry. The most important of these events was the entry of new investors in the oil industry in producing countries, whose aim was to secure permits for the exploration and extraction of 'cheap' oil. The arrival of these new investors was the culmination of two distinct trends in the consuming countries. The first was the emergence of national oil companies in the consuming countries with government support. The Italian company ENI was the first and most important case in point. The second trend was the growing emergence of independent companies, mostly American, which worked under a system of tax incentives in their own countries, like Occidental, Pan American, etc.

The first trend reflected the deep interest of the governments of consuming countries in the whole question of oil and its impact on economic growth. Many consuming governments in Western Europe (and also Japan) had a strong desire to gain access to 'cheap' oil outside the control of the multi-national oil companies, with a view to guaranteeing their own supplies. The most important factor driving these governments in this direction was the influence of the oil trade on the balance of payments, because Western Europe and Japan were forced, by the control exercised by the multi-national companies, to pay for their oil consumption in US dollars or, to a lesser extent, in pounds sterling.

10. In 1968 INOC concluded a 'service contract' type of arrangement with the French ERAP for the exploitation of certain acreage to the east of the Tigris. After the July 1968 revolution it started effectively exploiting directly the relinquished part of the Rumailah oil field (North Rumailah) with the technical help of the Soviet Union in accordance with the economic and technical cooperation agreement signed between the two governments in June 1969. The state-owned southern oil operations were on stream in April 1972.

The entry of their national oil companies into the investment field in the producing countries represented a saving of foreign exchange, as well as achieving some security over their oil requirements—which in itself provided a strategic goal. Thus, under the imaginative and courageous leadership of its President, Enrico Mattei, the Italian company ENI was formed as a new departure in oil-investment relations between the producing and the consuming countries and played a major role in breaking the stranglehold of the Cartel.

On the other hand, in the mid-fifties the United States set up a tax system devised to encourage American companies to invest in oil outside the United States, by creating tax incentives such as the 'depletion allowance'. This enabled companies or investment groups independent of the Cartel to penetrate certain producing areas with a view to obtaining permits for the exploration and extraction of crude oil. The aim of encouraging such American oil investment abroad was to sound out energy potential overseas. Energy policy makers of the US, aware of the fast rate of depletion of oil reserves since the mid-fifties, seemed to think that by encouraging oil investment abroad, they would secure for the United States increasing quantities of low cost foreign oil to keep pace with the oil import needs which the US economy might possibly have to face eventually.

The entry into the industry of these newcomers, whether national oil companies in Western Europe or independents in the United States, was made possible by devising new formulae for oil exploitation that differed from the conventional concession concept, particularly regarding financial terms. In appearance at least, some of these, such as 'joint ventures' agreements with the national oil companies of the producing countries, were more favourable than those of the concession system. This formula was based on investment costs, including taxes and royalties due to the host country, being shared equally between the foreign company and the national oil company, and equal sharing in the profit on the investment. The foreign company assumed the full investment risk in the event of failure, as against equal division of this risk with the national company in the event of success. As a result the 'joint venture company' paid to the government taxes and royalties at a rate of 50 per cent of the 'realized market' price per barrel less production costs and then divided the other half of the 'profits' equally between the foreign company and the national company.[11] The second formula adopted by the newcomers was the 'service contract', initiated by the French state-controlled company Entreprise de Recherche et d'Activitées Pétrolières (ERAP), by which the foreign investor acted as 'contractor' working in the producing area on behalf of the national company. The foreign investor did not enjoy the rights of ownership or 'equity share' in the oil venture and bore the full investment risks in the event of failure, with the national company bearing full costs of discovery in the event of success. Furthermore, the foreign investor secured marketing outlets for the national oil developed and produced from the areas covered by the contract. In return, he obtained secure access at low cost to a specific proportion of the crude oil produced.[12]

The entry of these new investors into regions dominated by the Cartel created a new situation that fed national

11. The net per barrel government-take was, in accordance with this formula, only marginally higher than the net per barrel host government-take under the conventional concession system. The reason was that taxes and royalties under the latter system were calculated in accordance with the posted price, whereas under the new system those taxes were computed on the basis of the realized prices, which in reality were lower than the posted prices, due to the various discounts that were offered to the buyers, especially during the sixties (see p. 52).
12. Under this system the oil reserves discovered by the 'contractor' were equally split between the 'national reserves' which were not covered by the contractual operations and were exploited by the national oil company at its option, and the reserves which were exploited by the contractor. Of the total volume of production from the latter, 30 per cent was secured to the contractor at a low cost. The method of calculating the cost of the contractor's oil was that 40 per cent of his share (or 12 per cent of the total production) was to be lifted at the production cost plus royalties (which at that time were 12½ per cent of the posted price) and 60 per cent of his share (or 18 per cent of the total production) was to be lifted at a price equivalent to the cost of production plus the royalties and taxes (which at that time were 50 per cent of the posted price). The remaining 70 per cent of the production from the reserves covered by the contractual operations was left to the national oil company to sell at prevailing prices in the world markets. In this respect the contractor was obliged to secure markets for the national company's share of production at a certain marketing fee.

aspirations to be rid of the multi-national companies' control and to resort to independent companies for securing technological expertise, crude oil marketing outlets, and capital needs to meet investment risks.

There were other more serious influences that contributed to shaking up the conventional concession system, especially the emergence of new producing areas with important oil potential which began to be exploited in new patterns. The chief characteristic of oil exploration in the new areas was the multitude of investors, and their mutual competition to secure permits to invest, in contrast to the traditional areas where one investor, the Cartel, exclusively controlled almost all the territory. In Libya, for example, there were about twenty-one companies holding permits to invest in specified areas. Despite the fact that a number of these investors came from the multi-national major oil companies themselves, they entered independently and not as partners with other multi-nationals in a consortium. The concession Esso obtained alone is a good example, as is the participation of BP with an independent American oil company, Bunker Hunt, beside a number of large independent companies, such as Occidental. The same, though on a lesser scale, applied in other new producing regions such as Nigeria.

This structural transformation of the investment system in the upstream phase of the international oil industry had serious effects, which were felt beyond this phase of the operations, especially in crude oil marketing, as will be seen in Part II. But what interests us here is that the emergence of the new investment patterns in crude oil production, independent of the major oil companies, inevitably led to free exchanges for the selling and buying of crude oil, outside the internal channels of closed circuit inter-company trade of the multi-nationals.

Clearly, as a 'free market' for crude oil grew, so the capability of the multi-national companies to control production of crude destined for international trade, and their planning of this production lessened. This led to an increase in the volume of oil being produced in search of independent market outlets, and so began an acceleration in the erosion of the power of the multi-nationals to strike a balance between the amount of crude oil supplies entering international trade with the amount being demanded. Further erosion in the control of the balance was caused by the governments of some producing countries exercising mounting pressure to increase their production, (as in Iran during the sixties). This meant additional quantities of crude oil becoming available, which exceeded the needs of the companies for their refineries and distribution networks, and this, in turn, led to an increase in the pressure of supply on the international market, and consequently weakening the companies' power to 'programme' production.

An end to the oil concession system

The developments reviewed above were among the most important factors which contributed to generating irreversible structural transformations in the international petroleum industry, especially in the system of exploitation of crude oil resources in the producing countries. Such transformations have resulted in the complete ending of the concession system and the replacing of the concessionary companies by governments of those countries in the ownership and management of the industry. In most cases this eventually led to complete government control, either by legislated nationalization, as in Iraq, Algeria, Venezuela and Libya, or as in other producing countries, by government participation in the concession through agreements with the oil companies, which led later to complete government take-over of the oil operations in return for guaranteeing the companies certain quantities of crude oil at a small price advantage in accordance with agreed long-term lifting arrangements.

In the case of nationalization, the existence of independent oil companies not only provided support for nationalistic aspirations for an independent system in oil exploitation in the producing areas, particularly in bearing the risks of investment, but also caused a major change in the nature of the world oil markets—an indication that the world oil markets could be penetrated. Unlike the conditions prevailing during the nationalization of Iranian oil, which made the marketing of nationalized oil practically unfeasible, the new emerging market conditions were such as to secure a minimum degree of success for the marketing of nationalized oil. Successful nationalization, limited though it was in the producing areas, began properly in Algeria, where some American independent companies holding small concessions were seized by the Algerian government in the wake of the June 1967 war in the Middle East. Despite the fact that the amount of production involved in the seizure was small in relation to total Algerian production, the success of the Algerian government in managing and marketing the oil helped it to penetrate world markets which, later on, enabled it to enter world markets on a wider scale. This more widely based entry followed subsequent nationalization measures covering the major part of fields exploited by the Compagnie Française des Pétroles, plus the cooperative arrangement between the Algerian National Oil Company (Sonatrach) and the French ERAP Company, so that today more than 80 per cent of Algerian oil is marketed directly by the national company.

In December 1971 Libya nationalized the BP share in the joint venture with Bunker Hunt. Politically motivated though it was (the implied British backing of Iran in its military occupation of small but strategic Arab isles in the Hermuz Straits in the Gulf), this partial measure of nationalization added a further impetus to the rising trend of fostering the states' rights to sovereignty over national resources and the take-over of oil operations.

However, the successful radical experiment in nationalization which completely reversed oil relationships in the Middle East, and from which stem all later developments in the system of investment, including the eventual total collapse of the system of concessions, was the nationalization by the Iraqi government of IPC (Iraq Petroleum Company Limited, covering the northern fields) in the summer of 1972. This was followed by the successful nationalization of the shares of the shareholding companies of the Basrah Petroleum Company (covering the southern fields) which in turn led by 1975 to total nationalization of all Iraqi oil.[13] As a result the production and marketing of Iraqi crude oil has come under total state planning and control through the Iraq National Oil Company and its subsidiary national companies and establishments.

But it must be emphasized here that what had taken place in Iraq prior to nationalization, in the way of direct exploitation by the Iraqi National Oil Company with Soviet technical assistance in the North Rumailah Field (recovered from the BPC Limited concession under Law No. 80 of 1961), had prepared the way for the successful nationalization of the large fields. Direct production and marketing of the oil from this oilfield by the INOC enabled Iraq to gain experience and market outlets, which later played an important part in overcoming obstacles to the marketing of the nationalized oil.

The nationalization of Iraqi oil led to tremendous pressure on the companies to accept such amendments in the system to meet national demands. Thus an important series of developments towards 'Participation' in concessions in the conventional areas of oil production was set in motion. In fact, the principle of government participation in the concessions has its roots in old agreements covering earlier concessions. The most important of these was the concession held by the Turkish Petroleum Company (later the Iraq Petroleum Company) in fulfilment of the San Remo Treaty which, in its day, recognized the right of the peoples of the oil producing regions to acquire a share with the foreign investors in oil operations and to own a specific proportion in those operations, defined in the Treaty at 20 per cent. But the way the concession of the Turkish Company was drafted, and the make-up of the company in law made it practically impossible at the time, in spite of repeated calls by the Iraqi government, to give effect to the article providing for government participation.[14]

13. Following the October 1973 Middle East War the American shares in the Basrah Petroleum Company (belonging to Esso and Mobil) as well as the Dutch share of Royal Dutch/Shell were nationalized. The Gulbenkian share and those of Britian (BP) and France (CFP) were nationalized in 1975.
14. The text of the agreement on participation was that the government of Iraq was entitled to 20 per cent of the share of the Turkish Petroleum Company

The principle of government participation was once again given stronger expression when OPEC, in its Declaratory Policy statement made in Resolution No. 90 of 1968, announced that it regarded participation as a fundamental principle in the case of states which chose neither nationalization nor direct investment in the exploitation of its oil resources. Following the patterns set in the IPC agreement, the Resolution also took the 20 per cent government participation ratio as a minimum base for any further negotiations with the companies. Later OPEC reiterated the principle of Participation by deciding to negotiate with the oil companies with a view to reaching acceptable formulae in the countries concerned.[15] In its resolution, OPEC set a basis for certain negotiating principles concerning the government share, the companies' compensation, etc. Accordingly, Saudi Arabia, with the support of OPEC (particularly those OPEC countries bordering the Gulf) negotiated with the companies.

The negotiations ended in signing the General Agreement on Participation between Aramco shareholders and the Saudi Arabian Government, effective from January 1973. It provided for state participation, which was to begin with a share of 25 per cent, to be maintained for five years and then to increase gradually every year until it reached 51 per cent in 1982. This agreement was the result of negotiations that were long and complex for important reasons, such as compensation of the companies for the installations, oilfields, and investments. The basis for compensa-

whenever those shares were offered to the public for underwriting. This never happened because of the nature of the company, being a consortium of shareholders with fixed shares and because of the strictly operational non-profit making nature of the holding company that had no shares to be offered to the public for underwriting.

15. The countries concerned were meant to be the member countries of OPEC bordering the Gulf, except Iran and Iraq also. Iran opted not to join OPEC negotiations on Participation because of the 1951 post-nationalization arrangements with the consortium, whereby NIOC became the owner of the oilfields and the concession holder, BP was paid indemnity in the form of compensation for the nationalization against the newly-established consortium operating on behalf of NIOC. Iraq did not participate in these arrangements either as it never entered into any agreement with the companies, not even with the Basrah Petroleum Company, which was left unnationalized for more than two years after the nationalization of the IPC in June 1972.

tion in the agreement was called at the time 'updated or revised book value', which virtually meant a revaluation of the obsolescent installations, by a calculation, in accordance with an index for the assets, that produced a new value several times the normal net book value. This also covered the disposal of the government share of crude oil and the definition of the respective roles of the government and the companies, especially in the area of marketing, in such a manner as to allow the government to gradually become an active partner in the management of the oil industry and ultimately to take a majority share.

But this agreement soon became outdated when the Kuwaiti Parliament rejected the participation pattern as laid down by the General Agreement on Participation, and when subsequently the government of Kuwait succeeded in reaching a new substantially different pattern of participation with the shareholders of operating companies. First, the Kuwaiti agreement gave, from the beginning, a much higher percentage to the state (60 per cent leaving the remaining 40 per cent of the oil to the companies to lift at the tax-paid cost). Further,company compensation was now to be based on normal net book values, *leaving* a much lighter financial burden on the government. The agreement also obliged the companies, if the government so opted, to 'buy back' partially or totally the government's share at normal market prices as determined by the government in the light of OPEC decisions. An even more important development occurred in Kuwait in 1975, by which the whole concession came to an end following an agreement with the ex-shareholders of the Kuwait Oil Company (BP and Gulf), whereby the government of Kuwait completely took over the concession and the entire operation of the oilfields covered by the foreign company, after converting the company into a totally national oil company. According to the new agreement, the ex-shareholders were compensated for the net book value against a long-term lifting arrangement (5 years, renewable for further five-year periods by mutual agreement) of certain quantities of crude oil which were fixed for the first agreement period at 40 per cent of the total allowable production (in the case of Kuwait 2 million barrels a day or 800,000 barrels per day for companies' liftings to be shared equally by BP and Gulf Oil). The sales contract provided a certain price advantage (15 cents per barrel) for the quantities lifted accordingly, in return for technological services to be rendered by the companies to the national oil company and other services such as transporting lifted crude on Kuwaiti tankers.

With the exception of the United Arab Emirates, which chose to keep the new participation ratio at 60:40, and the Kingdom of Saudi Arabia, whose negotiations with Aramco on a formula for a complete government take-over are not yet announced, the Kuwaiti formula is the one which prevails in practice in the Gulf area (except for Iraq, whose oil is totally nationalized), especially for the net per barrel financial results.

While a similar formula was applied to Iran, the financial results and the per barrel companies' margins on their liftings were under negotiation. However, the oil situation has completely changed following the Iranian Revolution, whereby the long-term lifting arrangements with the ex-shareholders of the consortium were put to an end.

All these formulae for government take-over turn on a fundamental point: that the state has become the owner and manager of all oil operations, including the exploration for oil, its transportation, exportation, and also marketing, either fully as in the case of nationalized oil or partially, as in the case of participation oil (with increasing government share). Secondly, all these formulae imply the assumption by the state of all the risks of investment in the oil sector, which means that the state has become responsible for the investment needed to develop discovered oil reserves, to search for new oil fields, or to increase the efficiency of oil recovery from the known fields. It also involved assumption by the state of all the capital costs necessary for development and construction work, for export, transportation and loading facilities, etc.

Consequence of state control of the oil industry in the producing countries

This fundamental change in the pattern of ownership of the

upstream phase of the oil industry in the producing countries has far-reaching consequences for the industry itself and for the economic return from it. It has, therefore, considerable significance for development in those countries, as their societies experience economic and technological change.

One of the most important consequences of this transformation is that the entry of the state into the planning and management of investment operations for the oil sector inevitably triggers other structural changes in the national economy towards closer integration of the oil sector into the national economy through intersectoral investment linkages, acquisition of technology, new employment and investment opportunities, etc. It also strengthens the link between the oil sector, as a source of energy or as an industrial input for the chemical, petrochemical and fertilizer industries or services, and the other sectors of the national economy. This could lead eventually to a substantial weakening of the 'dualistic' nature of the oil producing economies and, sooner or later, to an organic link between investment decisions on oil and other investment decisions related to national development programmes in general.

However, this process does not necessarily mean the complete and immediate integration of the oil sector in the national economy, since the dualistic nature of the economic structure will continue to exist, as long as the bulk of oil production is determined by exogenous factors, essentially related to the world oil markets, rather than by indigenous factors related to economic and social development.

Decisions on petroleum matters, including oil search and exploration, production capacity, export facilities, as well as decisions on pricing and marketing, etc. have, on their transfer to the state, become influenced by considerations of state sovereignty and its political requirements, especially those pertaining to economic growth and social development. This is important in defining the structural features of the oil industry at the production phase today because the oil sector is central to the national economy of the producing countries; in the final analysis it determines the level of economic activity, the foreign exchange reserves of the country, the general level of investment and employment, etc. A change such as we have described leads, naturally, to the integration of oil policy into the overall political and economic strategies of the state. The national oil companies which have replaced the foreign companies must, of necessity, operate within the framework of these strategies and their investment decisions will not be subject to purely commercial considerations and profitability criteria, but will be largely determined within the general framework of achieving economic and political development. In other words, the national oil companies, unlike the foreign companies, will not necessarily always act on the same principles as a commercial entrepreneur, especially where financial costs are concerned, since these now tend to be regarded as part of the general social cost of maximizing social welfare in the producing country.

Another important result of the government take-over of the oil operations in all producing areas, is that investment decisions in this sector are now limited to territorial borders and the sovereignty of the state, i.e. decisions in the oil industry are strictly confined within the framework of the national territory and become a part of the overall investment plans of the one state at the national level, in almost total isolation from oil investments and operations in other producing states in the same region. This is the opposite of the inter-territorial nature of the oil companies' operations, which in practice meant that as far as oil investments and operations were concerned, they were crossing the national borders of oil producing countries in which they were operating. In fact these companies' decisions were made in relation to investment decisions taken for other producing countries in an endeavour by the companies to achieve the highest possible economic return on their investments, not on a local basis (within one country), but on a regional and even international basis. The former pattern of 'inter-territoriality' in oil investment and operations has been replaced by a pattern of strict 'territoriality'. Thus, the previous linkages in the upstream investment system established by the companies in all the producing regions and especially in the Middle East, have been enormously weakened if not entirely broken in practice.

The impact of this change on the oil industry in the producing countries could be tremendous, as the nature of the strict concept of territoriality (excessive localism) prevents the realization of as high a profit return for the group of producing nations, as the companies used to achieve when they divided the investment risk among themselves on the basis of the region as a whole. The fact that today the state bears all the risks of investment, at a time when these financial risks are rising sharply, will make the cost borne by the state in the event of a failure very high indeed, and could adversely affect financial resources badly needed for the economic growth and social development of the nation as a whole. This is especially true in areas with impelling needs for extensive and costly social development programmes for raising general living standards.

Another important consequence of structural change is that investment decisions at the upstream phase of the oil industry are no longer linked with decisions and investments pertaining to the subsequent oil operations in the consuming countries (the downstream); this represents a change from the practice of the companies, who previously invested in crude on the basis of their forecasts for subsequent refining and distribution activities. The producing countries have begun to take their investment decisions from the point of view of a mere seller of crude oil, whose responsibilities are confined to delivering the oil to the main exportation terminals on his borders. In other words, crude oil production activities are not part of an integrated system for oil operations on an international scale, but have become investment plans formulated in accordance with purely national goals, and the most important of these is the need for financial resources to spend on economic and social development. In most cases such national goals are postulated in isolation from the size and nature of the markets for which the oil production is destined. These broken linkages in vertical integration between upstream and downstream and a general attitude of discouraging joint inter-OPEC regional planning could lead to real difficulties in coordinating production programmes and, *a fortiori*, joint regional planning of investment and production.

In fact, apart from problems caused by the management pattern of the oil industry in the producing countries, there are much more important factors inherent in the general structure of those countries that reduce the possibility of successful joint regional planning. Member countries of OPEC, as is well known, differ widely, especially in their economic and social structures, levels of economic development, absorptive capacity to spend on development (and hence financial requirements), population size, and other factors which determine expenditure requirements, such as armaments. Furthermore, oil producing countries differ enormously in oil potential as some countries have huge oil reserves but few other national resources that would allow for structural diversification in their national economy. This feature, especially if coupled with low population, as is the case in some of the Gulf oil producing countries in the Arabian peninsula, means a limited absorption capacity for expenditure on economic and social development. In contrast, other oil producing countries have very high absorption capacities for development spending and consequently greater need for financial resources, without, at the same time, enjoying great oil potential to enable them to expand in production capacity. This is why some of the oil producing countries are classified as 'surplus' countries with financial resources visibly in excess of expenditure requirements, and others are 'deficit' countries whose financial requirements exceed their oil potential. For the latter countries what matters most is to obtain as much per barrel income as possible through higher prices, whereas, for the former, the problem of oil prices must be dealt with in the light of their respective share in the future markets.

The lack of joint investment planning for the oil potential at the upstream phase of operations, and the almost non-feasibility of programming production of crude oil in the light of international market fluctuations, could have serious consequences for the oil industry both in balancing supply and demand, and strengthening of the oil price structure, as we shall see later.

PART II

The system for marketing crude oil: the fundamental structural transformations in world markets for crude oil

The vertically integrated nature of the oil industry (the organic linkages amongst the various phases) makes any changes in the crude oil extraction phase reflect on the ensuing phases. Thus the structural evolution observed in the discussion of the investment and exploitation system led to a radical change in the oil marketing system.

From the setting-up of the International Petroleum Cartel in the late twenties, to the aftermath of the Suez Crisis, the world oil market remained under the complete domination of the Cartel, mainly because of its controlling power over the sources of crude for export. Although to a much lesser extent, the Cartel also had control over the bulk of oil refining and transporting facilities in the world outside the US and the centrally planned economies. The horizontal and vertical integration of the major oil companies led, by virtue of this domination, to a special situation which became a characteristic feature of the international oil trade right up to the early 1960s; the international flow of crude oil was confined to a closed circuit of internal channels for the crude oil exchanges within the Cartel system of integrated operations. Hence a market, in the proper sense of the word, for crude oil entering international trade, did not exist. Indeed crude oil extracted and exported from the producing areas was not available for sale or purchase as such to Third Party dealers. Instead, it was exclusively traded internally among the companies, whether such crude was produced on account of each shareholder company of the consortium (lifted at cost) or exchanged according to long-term lifting arrangements between the crude 'surplus' companies like BP and the 'deficit' companies like CFP. However, both these two types of crude were produced in accordance with the same production plans of the consortia; this would make all crudes traded among the companies as a whole be kept within the same internal channels of the integrated operations, i.e. the entire quantities of crude oil produced were limited by the needs of the Cartel only. As a result there was no surplus of crude oil made available for sale to Third Party buyers, neither was 'deficit' crude made available for purchase by Third Party sellers. Sales or purchases made among companies, individually and through their subsidiaries and affiliates were of a purely book-keeping nature, mainly to serve tax purposes, and also to determine the profit margin for each affiliate, and the value added gained by the affiliates at each phase of the parent companies' operations. If, as an example, we take the Shell Company as illustrative of this feature, we find two types of crude oil made available to Shell—the first taken at cost plus from the producing region, where this company operates as a shareholder of consortia, in proportion to its share therein, and the second 'bought back' from the other 'surplus' shareholders in the consortia at the 'mid-way price'.[1] Shell then transports all the crudes that have been made available to it, on its tankers, which are either directly owned by it, or leased under long-term chartering arrangements, but operated by the same Shell 'shipping' subsidiaries. Likewise, Shell refines the crude, thus lifted and transported, in refineries that are owned and operated by subsidiaries sprung from the same parent Shell Company. Finally, refined products that belong to the same companies are distributed via networks run by other specialized independently operating Shell subsidiary companies. These subsidiary companies, belonging to the same parent company, handle the oil among themselves at the transportation, refining and distribution phases, at notional (or even phantom) prices at each phase, without there being any real dealing in the way

1. The 'mid-way price' is the price paid by the 'deficit' lifter equivalent at the cost of production plus half the difference between the cost and the posted price.

of buying or selling between the companies. These phantom costs for internal oil exchanges between subsidiaries of the same parent company ultimately determine the tax liability of the company's profits at each phase of the oil exchanges, and in every region in which Shell operates through hundreds of its subsidiaries. What is valid for the Shell company is valid in one form or another for the other companies in the Cartel.

This 'closed door' system of controlling inter-company oil marketing and trading operations was born of agreements which established the International Petroleum Cartel, particularly the Achnacarry Agreement, in the wake of intensive price wars that were waged by certain oil companies during the early 1920s, when each of them was working separately from the others. The report by the US Federal Trade Commission, published in 1952, gave a detailed and accurate account of how crude oil was handled within the closed circuit of the Seven Sisters (with CFP occupying a minor seat within the 'club') and of how their total monopoly of the international oil market meant that any attempt by other parties to gain access was doomed to failure. Any crude produced by a body outside the club, like for example a government in a producing country or an independent company, could not be sold to Third Parties because there were no channels through which such exchanges could take place. There was, therefore, no role for the governments of the producing countries in marketing oil operations since it was impossible—as proved by the nationalization experiment of the Mossadegh government in Iran—for governments to undertake investment at the crude phases without being able to guarantee a market for the product in which the investment had been made. Until recently this put the producing countries in no position to enter the oil industry and confined their role to that of collecting tax on the oil produced from their territories in accordance with the financial terms of the concessions (held in one way or another by those same major oil companies). The profit-sharing agreements concluded in the early fifties all spoke of the right of the state to take in kind, for its direct marketing disposal, a certain proportion of the oil produced in it, usually about 12.5 per cent of production (the royalty oil). Governments were in fact unable to opt for taking this share in kind, but preferred to take its financial equivalent computed on the basis of 'posted prices', because of the non-availability of valid independent marketing outlets for such crude.[2]

The emergence of a marginal 'free' market for crude oil

But the gradual cracking of the old system of oligopoly at the crude production phase led in its turn to a similar development in the structure of the world oil market, resulting in a weakening of the companies' control. One of the most significant developments of the fifties, and one which had a great influence on the nature of the market, was that economic difficulties—in particular those connected with the balance of payments in the post-war reconstruction of Europe and Japan—compelled the governments in some consumer areas to encourage investment in refining and distribution facilities at home independently from those controlled by the international Cartel. Such considerations were over and above any purely strategic considerations, which might encourage a move in this direction. One of the strategic targets for the construction of independent local refineries was security of supply of refined products depending on diversified sources of crude, concern for which grew after the interruption in the crude oil supply which followed the Suez Crisis. Although the capacities of those independent refineries were small when compared with the sum of existing facilities under control of the Cartel, their emergence nevertheless pointed to a new development in the international market, because they created an independent demand for crude oil. In other words they marked the beginnings of the emergence of oil

2. It is worth recalling that even in cases where outlets did exist, as for example during the sixties (see later text), there were still some financial restraints on governments opting to take their share in kind, because the financial equivalent of quantities lifted by the government was deducted from the government share on the basis of posted prices. This would make governments incur heavy financial losses every time they sold their share in kind at any market price lower than the posted price. This is why governments of the oil producing countries of the Middle East rarely took this option.

buyers seeking crude in a 'free' market in which the oil would be sold outside the channels of the closed internal circuits of the Cartel.

Another point: the independent oil companies, which had entered the field of investment in crude oil production under encouragement from the tax system and had started to produce crude from the producing regions, found themselves in a difficult marketing situation. From the marketing point of view, their investments were basically geared not to the international market as much as to the markets of their home countries. This was particularly the case of the American independent oil companies, who looked at the US markets as their main outlets for the oil produced from oil ventures outside the United States. But by the time the new production capacities came on stream, the American government policy on crude imports changed to closing the door to this independent oil, because the late fifties and early sixties saw growing restrictions on crude oil importation into the United States, and its confinement to a specified proportion of the consumption in order to protect domestic United States oil investments. As a result, the crude oil production of these independent oil companies, in search of immediate buyers, was offered on the international market, especially as the companies were not in a financial situation to allow themselves to wait long. In fact, because they had resorted to the international money market to finance the massive oil investments they had undertaken, it was in their interest to move their crude quickly in the international oil market, so as to generate sufficient funds to meet the bank loans and credits which formed the basis of the financing of their investments, and to offset their heavy cost.

Moreover, oil production in the Soviet Union was beginning to increase at a rather higher rate than its domestic consumption could absorb—owing to new oil finds in non-traditional areas, which led to Russian crude oil surpluses being available for export. The Soviet Union's desire to export crude was sharpened by its need for foreign exchange at a time when growth in the Soviet Union's domestic oil consumption was relatively slow, because of the then very high share of coal and natural gas in that country's energy structure.

This emergence of the Soviet Union added a new impetus to the formation of a 'free market' for buyers and sellers of crude oil independent of the Cartel. Indeed, the role of the Soviet Union was not confined to selling in a free market, but also to entering the market as a buyer in the late 1960s and early 1970s. Despite its export surplus of oil, quality balances in its oil requirements or economic advantages to be secured from transportation, made the Soviet Union buy oil from some producing regions and sell more of its own oil to the countries of Eastern Europe or the West.

The entry of national oil companies of the consuming countries in the investment phase of crude production in the producing areas was to secure oil imports for their own countries rather than to sell on the international market. Nevertheless, they ended in practice by partially entering the free market as both buyers and sellers at the same time. They did this in order to meet the requirement of the refineries that they owned, and to achieve quantitative and qualitative balances in these requirements. (The oil produced by these companies might be of a quality which did not suit the refining configurations or consumption patterns in their countries; this obliged them either to sell that oil and buy a more suitable type or to buy other oils for blending with the oil that they had secured).

All these developments led to a growth in the free oil market (in the sense of trading in crude oil independently from the major companies' closed circuits) and caused major changes in the world oil market and oil trade. Despite the fact that this 'free' market long remained rather marginal, never exceeding under any circumstances more than a small proportion (about 10 per cent) of the volume of oil entering world trade, its existence nevertheless marked the beginning of major changes in the market structure. Subsequent events led to an important curbing of the role of the major companies and the gradual entry into that market of the national oil companies of the OPEC producing countries.

The existence of this market was also a factor that strengthened the political trend in some of the producing countries towards a government take-over of the national oil wealth, because the presence of these marketing outlets guaranteeing a minimum level of government oil sales made the take-over a practical possibility.

On the other hand, the existence of this marginal market also weakened the inter-company relations in the Cartel, and led to competition among the multi-national major oil companies themselves. This was because some of the companies, notably British Petroleum and Esso, had begun to enter the free market as crude sellers, while others were doing so in a small way as crude buyers.

In more practical terms, it was the presence of that small but growing market, that has enabled the national oil companies in some OPEC member countries to gain, since the late 1960s, access to world oil marketing, albeit on a partial and limited scale. At that time the Algerian national oil company Sonatrach started to market the oil at its disposal from the production of the American companies of which the Algerian government had taken control following the 1967 war (see above). The Iranian National Oil Company (NIOC) was also able to market some small quantities of oil during the same period.

The growth of the 'free' market for OPEC oil

However, the crucial development that occurred at the beginning of the decade of the 70s radically altered the situation in the oil market and considerably strengthened the role of the national oil companies. These developments concerned both supply and demand. From the demand side, there was the increase in the entry of independent buyers into the world market, notably from the United States, the centrally planned economies, and the national oil companies of the industrialized consumer nations.

The entry of the United States into the 'free' market as a buyer of increasing quantities of crude, was an important development in the demand side of the market. As mentioned earlier, restrictions on importation of foreign oil were imposed in the past to protect US domestic oil production. However, the fast depletion rates of the reserves there, together with an increase in US oil consumption, led the United States government to reconsider its protectionist policy and resort to importing more oil; importation restrictions on crude oil were slackened. Thus, from being no more than 10–15 per cent of the total oil consumed in the United States during the sixties, the proportion of oil imported into the US has now reached about half of that country's total consumption (the quantities of oil currently being imported into the United States amount to over 8 million bpd).

The importation of an important part of the oil entering the United States is handled by channels independent of the major companies, since the number of refineries owned by the independents has grown considerably, and they have to enter the free market seeking crude oil in increasing quantities independent from the major oil companies. What has increased the importance of this feature, from the viewpoint of the producing countries from the Middle East and Africa, is the reduction in Canadian oil exports to the United States, which were not included in the importation quota system imposed on foreign oil entering the United States. The rapid growth of internal Canadian consumption and the halt in the expansion of its production (if not a continuing decline) led to a change in the status of that country from an exporter of crude to an importer. Obviously, such a change had to be reflected in larger crude oil exports from OPEC's Middle East and African countries to the Western hemisphere.

A significant development in the oil situation with the bloc of centrally planned economies has been the rapid increase in home oil consumption in the Soviet Union and the countries of Eastern Europe. These changes were long delayed owing to the energy structure in those parts being still largely based on coal and natural gas. Recent years have seen structural changes, with oil taking a larger share in the total energy requirements, particularly in Eastern Europe. The generally high economic growth rates in those countries, are also reflected in the increase in energy consumption, mainly oil. This situation has restricted the ability of the Soviet Union to increase oil exports, and led more and more of those nations to enter the 'free' market as buyers, particularly in the seventies. What gave impetus to the emergence of these countries as buyers of crude was their conclusion of broad agreements on economic and technical cooperation with a number of producing countries, such as Iraq, Algeria, and Iran. This created an additional demand for oil, either to serve as a means of payment for loans received by the producing countries under these agreements or for barter trading oil.[3]

The growing entry of OPEC national oil companies into world oil markets

The developments during the 1970s have been even more important for the supply side of independent oil entering international trade than the demand side, since they have led to a sudden and rapid increase in the quantities of independent oil being offered for sale by the oil-producing countries. In Algeria, the partial nationalization measures (51 per cent) applied to oil operations under the concessions, and the cooperative arrangements with the French companies, as well as the full nationalization of the small American companies which were put under government sequestration in 1967, led to the larger part of Algerian oil (80 per cent) being put at the disposal of the Algerian National Oil Company (Sonatrach) for direct marketing on the world market.

In Iraq, the success of the government in directly exploiting the North Rumailah oil field led to the successful entry of the Iraq National Oil Company into the international market as a crude seller, despite the legal obstacles raised by the foreign major companies (shareholders in the Basrah

3. This development in the energy situation in the Socialist Bloc has not, up to now, altered the nature of the Bloc as a net exporter of oil, for what the nations of this Bloc import in the way of OPEC oil is offset by Soviet oil exports to markets outside the Bloc (Western Europe). It is estimated that the latter position will continue for many years, although some reports, among them that of the Central Intelligence Agency, suggest that the position of the Socialist Bloc will be reversed in the early eighties and render its members net importers of oil and that will create additional demand for OPEC oil. This latter assessment of Soviet oil is, however, contested by other reports which predict that the Socialist Bloc will either continue to be a net exporter during the 1980s or at least self-sufficient.

Petroleum Company) in their allegations of rights to the oil extracted from the lands recovered under Law No. 80, to which reference has already been made. Shortly after the Iraqi company had successfully started marketing the oil exploited nationally, it entered the international market much more strongly in marketing nationalized oil after the promulgation in mid-1972 of the law for nationalizing of oil operations covered by the Iraq Petroleum Company Limited. The Iraq nationalization suddenly made available a quantity of crude exceeding one million barrels per day at the disposal of the Iraq National Oil Company for direct marketing. That volume was subsequently substantially increased following the successful ensuing nationalization measures taken for the shares of the Basrah Petroleum Company that led by 1975 to the full nationalization of that company. As a result, the entire production of Iraq is totally directly marketed by the Iraq National Oil Company.

The nationalization measures, especially those of Iraq, were indicators of the real structural change in the marketing relationships towards replacement of the major oil companies by the national oil companies of oil producing countries. These are gradually, but increasingly, penetrating a market that is no longer reflecting the domination of a few multi-nationals. On the other hand, nationalization means more self-assertion in fully implementing the concept of the producing states' right of inalienable sovereignty over national resources. Its real significance, therefore, lies not only in providing increasing volumes of crude for direct marketing by the producing countries, but rather in accelerating the qualitative transformations of the oil relationships. Despite the allegations of the companies, claiming rights to the oil being marketed by the national oil companies, and the consequent harrassment through the courts of a number of buyers of the nationalized oil, the national oil companies were never really impeded from penetrating the market in dealing with many independent or state-owned oil companies and refineries in the industrial countries. Those buyers who were practically the first to defy the multi-nationals, may have been aware since the beginning that the future of marketing relationships lies in direct dealings with the oil producing countries and their national oil companies. On the other hand, nationalization measures gave a real accelerating impetus for the oil companies, which had strongly resisted the principle of government participation (see Part I above), to accept and implement the concept in the oil operations covered by traditional concessions. This resulted in huge quantities of the government share of crude being available for direct marketing by the national oil companies.

Quantitatively, therefore, the implementation of the agreements on government participation played a very important part in increasing the volume of direct marketing by governments of oil producing countries. Those agreements suddenly made available increasing quantities of that crude, should the producing countries so choose. More importantly, subsequent developments in the concept of government participation and its practical implementation led to a substantial increase in those crudes entering effectively the world markets.

At the start of the Participation Agreements, the quantities of oil available for direct marketing by the government were limited. In addition to the low ratio of participation, which was specified at the beginning as 25 per cent, to last for five years before it would start gradually increasing to 51 per cent only, another five years later, the agreed lifting arrangements with the companies had resulted in the bulk of the crude owned by the governments being sold back to the companies in the form of crude buy-back arrangements. According to the initial provisions of the Participation Agreements, producing governments accepting that pattern of marketing relationships were under obligation to resell to the companies a proportion of their share, which was decreased annually until it vanished after three years (the Bridging oil).[4] At the same time the companies were obliged, if the government so chose, to buy back an increas-

4. During the initial period of the agreements' implementation, Bridging oil amounted for the first year to 75 per cent of the government share to be decreased during the second year to 50 per cent and then to 25 per cent during the third year, after which this form of Buy-back vanishes completely.

ing part of the government's share, (the Phase-in oil).[5] One of the objectives of these arrangements was to bring about the reversion of the major part of the government share in crude to the companies, to meet their obligations to buyers with whom they were linked by contracts, and to grant them a time period in which to fulfil these obligations (this applying to the Bridging oil). The other aim was to prevent large quantities of oil, marketed directly by the governments of the producing countries, from suddenly entering the world market with huge quantities which, so it was said, would disturb it, (this applying to Phase-in oil). So the entry of the national oil companies of these countries into the international market was intended to be made gradually. In respect of these two types of buy-back oil, the companies enjoyed discounts which were initially very large indeed and brought them considerable profits,[6] particularly with the sharp rising trends in the prices realized on the international market, as in 1973. What is more important, as far as marketing is concerned, is that those arrangements substantially diminished the amount remaining to the governments for direct governmental marketing, so that initially it did not exceed 10 per cent of the government's share.

However, those government/companies marketing relationships proved to be short-lived and have later been drastically altered so that larger quantities of crude oil marketed directly by the government have been increasingly

5. According to the lifting arrangements of the General Agreement on Participation, Phase-in oil (which was an obligation on the companies and an option for the government) accounted for 90 per cent of the government's share during the first year (including of course the quantities of the Bridging oil), gradually decreasing to reach very low percentages of government share, then to vanish completely after 10 years for the initial 25 per cent government share, and 9 years later for the increments of that share.
6. According to the initial agreements on Buy-back oil, the cost price paid by the companies on Bridging oil was equivalent to the Quarter Way Price plus an amount fixed at about 10 US cents per barrel, with the intention of bringing the price near the going market price (the QWP means the tax-paid cost plus one fourth of the difference between the tax-paid cost and the posted price). On the other hand, the companies paid for the Phase-in oil at a price equivalent to the tax-paid cost plus 19 US cents per barrel, with the intention of bringing the total price up to the market price less an 'advantage' to the companies.

appearing in the market. In Kuwait the rejection of the participation pattern as defined in the General Agreement on Participation, (see page 25 above) and the success of the government in achieving a new pattern for participation has strongly contributed in weakening the role of the major oil companies in the marketing of crude produced by the OPEC countries, especially those of the Gulf area. The increase in the government share to 60 per cent in accordance with the new pattern of government participation and the absence of a government obligation to resell, partly or totally, its share to the companies in the form of Buy-back oil, practically and ultimately reduced the amounts of crude oil lifted by the major oil companies to their participation share, i.e. only 40 per cent. Again in Kuwait, the further change in government/company relationships that occurred through the conclusion with the companies the agreement on full government ownership, i.e. the take-over arrangements, had given further impetus to the rising trend of direct marketing of crude oil by the national oil companies of the oil producing countries. This was in spite of the fact that more or less the same ratio of lifting by the ex-shareholders of the concession, i.e. 40 per cent, was still maintained by the new arrangements, albeit with more advantageous financial results for the government. The new pattern implied greater involvement of the governments in the oil industry and the establishment of new state-owned institutions for the management of the oil operations and the government supervision thereof. This necessarily had the effect of making available greater quantities of oil for direct marketing by the national oil companies, since the entry of the government in these new areas of activities would create greater incentives for the national oil companies to be more active on the international market.

The formula of government take-over by agreement with the companies had a marked effect on international marketing relationships in the Gulf area. In Qatar there was a full government take-over in accordance with the same new pattern (with a few amendments). The United Arab Emirates, which chose to keep the older pattern of government participation on the 60:40 per cent basis, without implementing the new pattern of complete take-over, was nevertheless successful in directly marketing the bulk of the government share through its national oil company.

Iran, on the other hand, earlier adopted another system of government/company marketing relationship, although the same per barrel financial results of the participation formula was maintained. As described above, Iran did not adopt the formulae of Participation because, following the nationalization by Mossadegh, the system of oil exploitation replacing the old concession, was based on the concept of keeping the national oil company of Iran as the owner of the nationalized oilfields. From the purely legal point of view, the consortium of foreign companies exploiting the Iranian oil worked on behalf of the national oil company with the proviso that the net per barrel government take in Iran would be the same as the other producing countries' in the Gulf area. According to those arrangements, the compensation against nationalization was already paid by the Iranian Government at that time. This was why Iran chose to adopt the formula of government take-over earlier than others instead of adopting the Participation formula. The ex-shareholders of the Iranian consortium became lifters of crude oil on the basis of long-term lifting arrangements (25 years) by which the companies enjoyed price discounts against lifting oil in a manner which made the per barrel net government take in Iran equivalent to that in the other producing countries of the Gulf area which chose Participation. This financial arrangement was subject to revision, every time there was a change elsewhere in the area in the financial relationship of oil lifting in accordance with Participation arrangements. As for the quantities of oil marketed by the National Iranian Oil Company, they were initially modest but increased gradually in accordance with the same agreement until 1978 when they accounted for about 30 per cent of Iranian production. Besides these lifting arrangements another technical government/company relationship was created for the management of oil operations in Iran whereby a new company was established with the technical aid being part of the total take-over package.

However, the situation in Iran has been completely revised after the Iranian Revolution, subsequent developments have put an end to all those long-term lifting arrangements with the ex-shareholders of the consortium. As a result all Iranian oil is currently marketed directly by the National Iranian Oil Company on the basis of sales contracts that are applied to all buyers taking Iranian oil at a market price fixed by the government in accordance with OPEC resolutions. Buyers include some of the ex-shareholders of the Iranian consortium, whose current purchases from Iran have been enormously reduced, and account for a marginal proportion of the total Iranian production.

In Venezuela the ownership and management of the oil industry has reverted to the government in accordance with the promulgation of the 1974 Law of Nationalizing of Venezuela Oil Operations that was put into effect the following year. However, negotiations were conducted with the ex-shareholders, mainly Exxon and Shell, which led to some arrangements by which those companies continued to lift an important part of Venezuelan oil, currently two thirds, under conditions which are similar to those applicable in Kuwait, i.e. the companies enjoy a 15 US cents advantage on the price against the technological services they render the national oil company.[7]

As mentioned above, the results of negotiations with Aramco with a view to finalizing a complete government take-over are yet to be announced. Meanwhile, the successive changes in the government/company relations still apply—at least in so far as the per barrel financial results for the government are concerned. The amounts of oil marketed directly by the Saudi National Oil Company (Petromin) are still limited owing to the continuation of many of the marketing arrangements (namely percentages of Buy-back oil) which were agreed upon in accordance with the former pattern, in that the ex-shareholders in Aramco still lift the bulk of the oil produced in Saudi Arabia, leaving to

7. The national oil company in Venezuela, Petroleos de Venezuela, in fact represents a holding company responsible for corporate finance and planning as well as marketing. The subsidiary companies (Lagoven, Maroven, etc.) are merely operational companies.

Petromin's direct selling the lesser part of the government's share. Initially, Petromin's share for the direct marketing was about 5 per cent of Saudi Arabia's total production, in implementation of the same ratios stipulated in the General Agreement for Participation (after deducting both Bridging and Phase-in oil). The remainder was lifted by Aramco's shareholders. Recently however Petromin's share was increased noticeably. Of the total 8.5 million b/d which is the current national ceiling for Saudi Arabian production, Aramco shareholders are currently lifting less than 7 million b/d or about 80 per cent of total production, with Petromin taking the internal and external marketing of the remainder.[8]

The fading power of the multi-nationals in world markets

Despite these large differences between countries within OPEC in respect of the proportion of oil marketed directly by the state—in some cases amounting to all the oil produced and in others to only a small part of the production—nevertheless these developments in the system of marketing OPEC oil have all led to the sum total of oil marketed directly by the member countries of OPEC being equal to about half their total production. This included the full nationalization of oil, with the national oil company marketing all the nationalized oil (Iraq), or part of it (Venezuela), partial nationalization (Algeria and the Libyan Jamahiriya), or Participation (UAE), or government takeover by agreement (Kuwait). The growing role of the national oil companies in the area of marketing is naturally accompanied by a growing power in the world market. In fact there has been a continuing shift in the balance of marketing power, started rather forcefully by nationalization measures especially in Iraq, which initiated direct market relations between producers and consumers (see p. 17 above). Recently, however, in the wake of the events that led to the collapse of long-term marketing arrangements in Iran, along with other marketing developments like

8. Current external direct marketing by Petromin amounts to 1.2 million barrels per day, or about 15 per cent of the allowable national production ceiling of Saudi Arabia.

nationalization of BP in Nigeria, the trend for a greater role for the national oil companies in controlling the market and actually in setting prices, has been strengthened.

The current marketing situation of the multi-national major oil companies shows a dramatic reversal in the relationship between the quantities of crude oil to which those companies have access and the quantities of crude oil which are processed by them in their own downstream operations. As recently as the first half of the 1970s, those companies were disposing of crude oil much in excess of their own downstream requirements, so that they were net sellers of crude in substantial quantities in the world market. In 1975, for example, the total oil production of crude accessible to these companies (through Participation, lifting arrangement, etc.) was about 25 million barrels a day, against total crude processed by them of about 19 million barrels a day, thus leaving a gap of about 6 million barrels a day of crude oil, or about one third of the companies' crude processing requirements, to be offered for sale as crude on the international market. Obviously this situation enabled the companies to control the market and to lead the setting of oil prices in that market to levels which other sellers, including the national oil companies of the producing countries, had to follow. The dramatic reversal in the companies marketing and crude processing ratios has been such as to shift completely their status from net sellers to net buyers. Against the companies current crude requirements for processing 20 million b/d, their access to crude oil lifted in accordance with Participation or take-over requirements will not exceed 18 million b/d, thus creating for them a deficit of about 2 million b/d of crude to be filled by buying from OPEC's NOCs to meet their processing requirements. This situation has actually deprived the companies of their controlling power in the market of crude oil. The controlling power now lies with the national oil companies who are increasingly assuming the role of market leaders, including the setting of the 'market' price, in accordance with OPEC price decisions.

The OPEC power in the world market: some marketing issues

This growth of the role of the state in direct marketing is linked to the strengthening of the trend of extending the states' sovereignty over its oil resources and its management role of the oil industry. It also implies a higher degree of integration of the oil industry into the national economy, thus creating in its turn radical changes in national economic and social structures.

However, the great disparities that exist in the pattern of government control in the industry and the fact that governments of the oil producing countries are acting individually and in almost complete isolation of each other in the area of marketing and investment planning, could lead to serious consequences for this industry in general and oil marketing in particular, especially when the market is depressed as a result of stagnant demand, as was the case from 1975 through to the first half of 1978.

For example, one result of the Participation Agreements, and the following government take-over by mutual agreement with the oil companies is that the companies are in a better relative competitive position, as far as the terms of crude lifting arrangements are concerned, because these arrangements secure for the companies a profit margin that ranges between 15 cents per barrel as a minimum, for the long-term purchase contracts in the Gulf area, as in the case of Kuwaiti oil, to more than one dollar per barrel for the crude liftings from its 'equity' share in Participation, depending on the terms of lifting arrangements, participation patterns, and production costs.[9] It should be stated, however, that this situation represents a very great improvement compared with the huge profit margins that the companies were enjoying in accordance with previous lifting arrangements and patterns prior to the latest Participation or take-over agreements. The reason was that Par-

9. On an average OPEC price of $20 per barrel, the tax-paid cost should be around $19 per barrel (88 per cent of a computed post price for tax purposes—1.075 of the OPEC price or $21.5 per barrel = $18.92 per barrel plus the companies' share of the cost of production.)

ticipation provided a lifting cost on the companies from their Equity oil (their share of Participation) equivalent to the tax-paid cost (which meant the companies' share of the production costs plus taxes and royalties paid to the government) whereas revenue from the government share, whether through direct selling of its share in the market or through the Buy-back arrangements, was based on the official OPEC market price (with the government bearing its share in the production costs). Therefore, when the tax was 55 per cent and the royalty 12.5 per cent of the posted price, the per barrel tax-paid cost was of the order of 60 per cent of the posted price plus 45 per cent of the production costs.[10] As for the price of the Buy-back oil (the government's share sold back to the companies) it was generally set at 93 per cent of the posted price.[11] According to this rate, therefore, the cost of the companies' share of the oil (the Equity oil) was less than the cost of Buy-back oil by more than 32 per cent of the posted price. On the basis of the then prevailing posted price for Arabian Light, amounting to $10.84 per barrel (by middle of 1974), the companies' per barrel tax-paid cost was marginally over $7 against a price of over $10 per barrel for the Buy-back oil. The companies' per barrel tax-profit margin from selling its Equity oil on the free market amounted to about $3 per barrel. However, the average per barrel cost of the total oil lifted by those companies (assuming that the complete government share was also lifted by the companies in the form of Buy-back oil at 93 per cent of posted prices), was less than 70 per cent of the posted price, i.e. a profit margin of about 25 per cent of the posted price.[12]

10. The per barrel net government take = $0.125P + 0.55\ (P - 0.125P - c)$
The per barrel tax-paid cost = $0.60625P + .45c$
Whereby P is posted price and c is production cost.
11. The ratio of 93 per cent of the posted price was taken from Petromin's sales in the middle of 1973 of the remainder of the Saudi government's participation share (after deducting Bridging and Phase-in oil) to some independent, mainly Japanese, companies. Later on that ratio was generally applied as an indicator of the market prices for the Buy-back oil.
12. The profit margin mentioned was calculated on a state participation of 25 per cent (in accordance with Participation arrangements of the middle of 1974, and before its amendment by Kuwait), considering that 75 per cent was taken at the tax-paid cost, which was equivalent to 60 per cent of the posted price,

The major companies were thus placed in a very strong competitive position compared with the independents, who bought oil at a higher price (the official OPEC price) and this led to the major companies realizing huge unearned windfall profits. As a result, OPEC passed various resolutions to reduce the size of the companies' profit margin by increasing the rate of tax and royalty in a number of stages. The last of these OPEC decisions on tax and royalties was taken in November 1974 at the Abu Dhabi meeting of OPEC members bordering the Gulf when the tax rate was raised to 85 per cent of the posted prices and the royalty to 20 per cent,[13] thus raising the per barrel tax-paid cost of the companies' liftings from their Equity oil to 88 per cent of the posted price.[14] In other words, the companies' share from the oil was lifted at a cost of only 5 per cent of the posted price less than the cost of the sales on Buy-back. To put it another way, in relation to the sum of the oil lifted by the companies, the average cost to them per barrel rose to 91 per cent of the posted price, implying a reduction in the profit margins of the companies down to only 2 per cent of the posted price.[15] By this reduction in

whereas the government share (25 per cent) was sold back to the companies at a price equal to 93 per cent of the posted price $(0.75 \times 0.6P) + (0.25 \times 0.93P) = 0.6825P$, taking the companies' share of the cost into account. It must be pointed out here that this profit margin was substantially reduced after the amendment of the tax system by increasing the rate for royalty and tax to 20 per cent and 85 per cent respectively, as will be stated in the text.

13. The amendments to the tax system ran as follows:
The royalty rate was increased from 12.5 per cent of the posted price to 14.5 per cent in June 1974 by resolution of the OPEC 40th Meeting of the Conference. At its 41st Meeting (September 1974) the Conference decided to increase the tax rate from 55 per cent to 65 per cent and the royalty rate from 14.5 per cent to 16.6 per cent of the posted prices; finally, the tax rate was raised to 85 per cent and the royalty rate to 20 per cent in November 1974 by the member states bordering the Gulf, at a special meeting held in Abu Dhabi, which was endorsed by OPEC at its 42nd Meeting (December 1974).
14. The per barrel government-take:
$= 0.20P + 0.85\,(P - 0.20P - c)$
$= 0.20P + 0.85\,(0.80P - c)$
$= 0.88P - 0.85c$
15. Considering that 40 per cent of oil was lifted by the companies at the tax-paid cost, (88 per cent of the posted price) against lifting of the remainder in the form of Buy-back sales at a price equivalent to 93 per cent of the posted price $(0.4 \times 0.88P + 0.6 \times 0.93P = 0.91P)$.

their profit, the companies found their windfall profits tremendously reduced.

Yet, despite this reduction in the companies' profit margins, the price advantage that they enjoyed still placed them in a favourable competitive position vis-à-vis the national oil companies which sell oil directly at the official prices in accordance with OPEC decisions. This distinctive competitive position of the major companies had the effect of weakening the independent oil-purchasing companies, or the national oil companies in the consumer countries, which were obliged to buy oil without any preferential treatment. Their position as principal outlets for free oil, i.e. oil marketed independently from the major oil companies, could be weakened, especially since the financial status of most of these companies was not strong enough to support their standing for long in the way of the financially stronger giant companies. This situation affected the national oil companies of producing countries in two ways: first, a weakening of their bargaining position in concluding agreements with Third Party buyers, and in consequence an impediment to expand and secure greater control of the oil industry (and clearly this result had greater effect on the producing state whenever the proportion of oil directly marketed in this state increased); and second, in the event of a market slump and reduction in demand, a threat to the volume of their sales and consequently to their share of the market, should they be obliged to keep the official price.

However, this situation of relative competitive advantage of the major oil companies over the national oil companies has lost most of its significance due to recent developments which completely changed the balance of power in controlling the markets in favour of the national oil companies, especially with the major companies becoming net buyers, or at best self-sufficient in crude.

On the other hand, the fact that the national oil companies come to sell crude oil on the free market without owning the means of controlling the market outlets, in the sense of ownership of refining and distribution capacities, could expose them to intense competition amongst themselves, especially in a period of market 'glut' when the buyer is in a stronger bargaining position. Unlike the major oil companies, which resorted to long-term production programming based on their own forecasts for their down-stream operations, the producing governments found it difficult to agree on production programming, so as to guarantee the necessary balances in world supply and demand. Despite previous attempts to programme production, OPEC, as is well known, has up until now failed to create a workable system for collective control of production so as to strengthen the sellers' position at times when there is a market 'glut' or stagnant demand. Competition among national oil companies to secure a larger share in a shrinking market could expose the price structure to the dangers of downward pressures.[16]

Numerous attempts have been made in the past, and within OPEC, to establish some sort of coordination among national oil companies, particularly for marketing. Some of the earlier decisions (since the sixties) taken in OPEC were aimed at setting-up a type of institutional agreement for this purpose.[17] Serious attempts were also made in the seventies to create some channels for the coordination of marketing operations and even for handling some of the

16. It is obvious that the current situation of the oil industry which is characterized by strains on supply and market shortages since the last quarter of 1978, will replace those downward pressures with upward ones, which would remove any danger that could result from the lack of market coordination among national oil companies. It is not unlikely, however, that such downward pressures on the price structure would reappear as soon as the current state of market shortages disappear and are replaced by a slack market.
17. Between 1968 and 1971 a number of meetings for the representatives of national oil companies of OPEC were held to discuss certain forms of coordination among them, especially in the area of marketing, in order to avoid harmful competition. Certain ambitious ideas, such as the creation of an association of NOCs in OPEC countries, as was suggested by some consulting firms, was completely ruled out by the NOC representatives, who were satisfied with such more modest arrangements as establishing a department within the OPEC Secretariat for coordination between the NOCs. However, the OPEC Conference was not in favour of such arrangements, especially as there were some fears within OPEC that any institutional arrangement for the NOCs would affect the mechanics of central policy-making decisions within the 'umbrella' of OPEC. Furthermore, there were some other fears, unjustified in fact, that institutionalizing the coordination among NOCs might lead to an overlapping of functions within OPEC. Consequently all these meetings were finally halted and endeavours in this direction shelved.

price problems such as relative values, but all have not been completely successful.[18]

Recently, however, meetings of the OPEC national oil companies' representatives are regularly held at the OPEC Headquarters to exchange views on certain possible areas of cooperation, such as technical cooperation, manpower planning and development, etc.

18. In the summer of 1974, a meeting of the NOCs' representatives responsible for marketing was held in London to examine possible ways and means for cooperation among them at a period when it started to be very clear that the market was slackening and that already there were some signs of market reversal to the side of buyers. A document calling for some forms of cooperation was actually signed, but ratified by a limited number of NOCs, which made it a dead letter, and this again led to the shelving of the idea of cooperation among OPEC NOCs.

PART III

The system for pricing crude oil exported from the oil producing countries: the changing power relationships in pricing raw materials

Developments that have taken place in the field of oil pricing during the decade of the 70s can perhaps be regarded as the most important transformation in the structure of the world petroleum industry. They have led to a real alteration in the balance of world economic power, in a manner that has brought about a more equitable distribution of economic growth generated from oil between its owners and its consumers.

Among the most significant features of the colonial domination exercised by the Great Powers over the oil producing regions was the system for pricing oil imposed on those countries by the members of the International Petroleum Cartel. It determined the share of the producing countries in their exhaustible natural wealth. As in the case of the system for oil investment and exploitation (regarding oil search and production), the pricing system reflected—perhaps even more flagrantly—the extent to which the governments of producing countries were deprived of exercising their right of sovereignty over their natural resources. The concession system had effectively negated this right transferring it in reality to the companies, which fixed price levels unilaterally. Prices were set not to find the true value for a commodity entering the international market—as we have explained, there was no such market—but for the purpose of fixing income tax and royalties, i.e. the governments' share of the oil wealth in accordance with what was termed the posted prices. Those prices were published by the selfsame companies in the specialized press, the most important being *Platt's Oilgram*. The system for posted prices began to show its particular importance for the oil producing countries when the so-called profit-sharing formula of the concessionaire system first appeared in the early fifties; under the formula, the profit from oil operations in the producing countries came to be defined as the difference between the posted price at a sea terminal (if this terminal lay within the territorial borders of that country) and the cost of producing the oil and getting it to that terminal, such that the government's share was defined as half that difference. It was regarded at one and the same time as a compensation in the way of income tax and royalty paid to the owner of the land for its exploitation.[1]

The story of oil prices in the Middle East has highlighted the imbalances that characterized the system of international economic relations, which was tailored to serve the interests of the industrialized countries to the detriment of those countries that were producers and exporters of raw materials. These imbalances represented a true squandering of the resources of the oil producing nations.

The unreal systems for pricing oil that, for a long time, the oil companies were able to implement—though only with the support and protection of the colonial powers—kept the share of the oil producing countries at artificially extremely low levels, so low that the real cost of acquiring oil by the Western developed economies was almost nil. This reduced the pricing system to the level of plunder, particularly since the ultimate share of the producing countries was caused by the concession system to be less than half the posted price.[2] These very low levels in the

1. Before this system was brought into effect posted price played no part in setting the governments' share of the exported oil, it being set as a fixed sum per ton lifted (4 shillings—in gold—later 6 shillings per ton).
2. As a result of the last cut unilaterally decided by the companies in the posted prices for crude oils exported from the Middle East Gulf area in September 1960, the governments' per barrel take dropped to an average of 80–90 US cents. The posted price of Arabian Light, for example, fell to $1.80 per barrel, which meant, according to the provisions of the concession system, a government-take equivalent to 85 cents on the assumption that production

cost of acquiring oil by the consumer in the industrialized countries can be considered as one of the main factors for the success of these countries in achieving high rates of economic growth and technological advancement. The pricing system thus constituted a veritable subsidy for the economic growth of the industrialized countries that the producing countries had to transfer under this imbalanced system of international economic relations.

The facets and forms of this order of pricing were created by the International Petroleum Cartel. The most important of these concerned the companies' domination of all crude oil sources in the exporting areas and their power to plan crude production in those areas in the light of their downstream operations, in a manner that ruled out any independent exchanges of crude oil. Such a closed system for producing and marketing oil enabled the companies to create a system for the pricing of oil, which was in effect no more than a series of decisions taken by them for their own ends, without those decisions being motivated by economic forces, such as supply and demand especially in the long-term.[3] Such pricing decisions of the companies had no real economic significance simply because channels for the sale and purchase of crude oil outside the companies' integrated operations did not exist, except margi-

costs were equal to 10 cents per barrel. The government-take stayed at this level until early 1971, when it was increased to about $1.05 by the middle of February of that year, the effective date of the Tehran Agreement.

3. It is very interesting to compare the movement of the posted prices with the relative movement of reserves and production (the additions to reserves versus the net depletion from reserves, or what is called the reserves to output ratio), together with that of demand for OPEC oil. The sixties witnessed a regular and steep decline of the Middle East reserves' ratio and a regular and steep rise in world demand for OPEC oil, a situation indicating a growing scarcity of oil, in the sense that the trend of adding to reserves is slower than the trend of depleting them. In similar situations, economic theory suggests that prices should increase as the reserves ratios decline and demand grows. The companies' pricing policies were to cut down Middle East oil prices! It was only following the OPEC intervention that oil prices were frozen in dollar terms, hence declining in real terms. The irrationality of the companies' pricing policies are well shown in the diagram below. It was only after the government take-over of the OPEC price that price movements began to indicate economic rationalization. Rising prices are a reflection of growing scarcity.

nally. For the companies, the real significance for the pricing of oil was not for the determination of their profits at the stage of lifting crude oil, as the companies' real profitability was measured by the global financial results of their world-wide operations.[4] Its significance lies rather in determining the cost of lifting crude oil, including taxes and royalties paid to host governments, but only as one component part of the global costs of the integrated operations. Hence the fictitious nature of what was at that time called the profit-sharing formula between the host countries and the concession-holding companies.

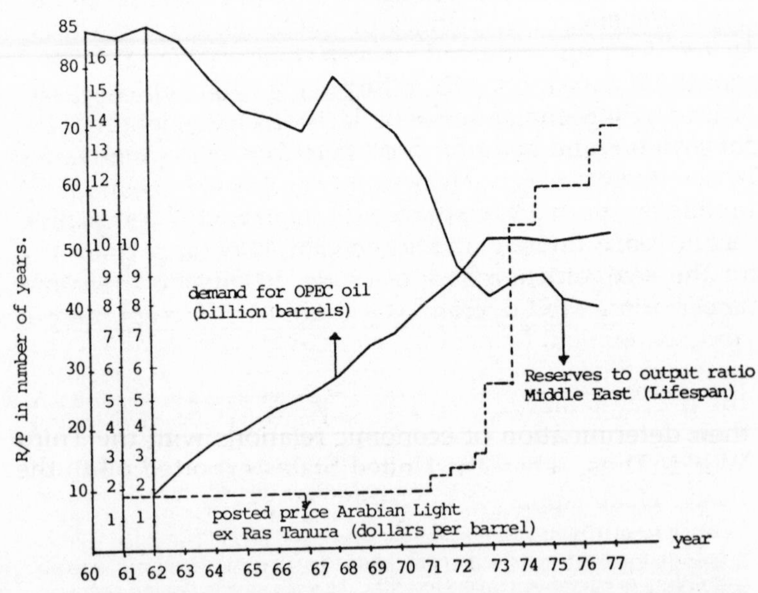

4. The corporate profits of the companies are the difference between the revenue from the sale of refined products all over the world and the total costs incurred in carrying out all the integrated, world-wide operations. Each phase of the operation before the sale to the final consumer is a cost to be added to the global costs that ultimately determine the global corporate profits. Hence, the taxes and royalties paid to the host countries are no more than costs to be added to the other costs of the integrated operations. In this sense, therefore, posted prices are no more than a means of determining the costs incurred by the companies at the upstream phase of their integrated operations.

Before the birth of this system, decisions taken by the companies on oil pricing in the Middle East served purposes which had no connection with the determination of the producing country's share, since that share was determined in accordance with the former concession system—flat royalties per exported ton—but which were connected with the cost of the inter-company transfer of the crude oil via their internal channels (such as the cost of transferring the crude oil from companies in surplus to companies in deficit on the basis of the inter-company pricing system—the half-way price). Moreover, posted prices served as a notional basis for determining the book cost of oil dealings among the subsidiary companies of the same parent companies. In addition, the posted prices system served as an umbrella for the determination of the true costs incurred by the consumer countries in acquiring crude oil through world trade. For consuming countries with access to concession oil, the cost transfer took the form of royalties and taxes paid to the producing countries. For those countries without such access, posted prices served as a framework for international payments against their oil acquisitions through the major companies.

The companies would not have been able to impose artificial prices in isolation from the international political and economic relationships reflecting various balances in power politics. On the contrary, oil prices always reflected the overall political strategies of the Western countries in their determination of economic relations with the Third World. Thus, when the United States exported oil to the Eastern Hemisphere in the period before the Second World War, the companies fixed the oil price in the Middle East at a level equivalent to the oil price in the Gulf of Mexico for similar oil, plus the cost of transporting it from the Gulf of Mexico to the point of destination. At that time this was called the Single Basing-Point System or US Gulf Plus. For a purchaser in the Eastern Hemisphere buying Middle East oil, the addition of these 'phantom' transportation costs rendered that oil costlier in practice than oil from the United States. The objective of this unreal pricing system was to protect the price of US oil exported to Europe, at a time when Middle East oil did not account for a high proportion of European oil needs. The share of oil in total energy consumption in Europe was very low, because at that time the region still depended on coal as its main source of energy.

However, developments in the oil industry during and after the Second World War radically changed these oil relations, mainly because the United States changed its status from a net oil exporter before the war to a net oil importer after the war.

During the war, under pressure from the Allies to reduce the cost of oil necessary for the marine war machine, especially in the Mediterranean and East of Suez, the companies replaced this fictitious pricing system by another, no less fictitious, called the 'Double Basing Point'. This involved establishing a second basing point in the Middle East (the Gulf) to match the first basing point in the Gulf of Mexico; in other words, the equalization of the price of the Middle East oil and that of the United States oil, at the export terminal (FOB). The concept of phantom transportation charges was thus done away with, and this led to the first significant reduction of Middle East oil prices for the European importer.[5]

The United States' involvement (the Marshall Plan) with the reconstruction of Western Europe (and Japan), by providing the necessary cheap energy to generate the technological changes required for this reconstruction, led the West to pursue policies designed to lower the cost of Europe's acquisition of Middle East oil, particularly since the United States had ceased to be an exporter of oil to the European markets. The companies played a major role in coordinating with the architects of Western economic policies, particularly the European Cooperation Administration, in successfully lowering Middle East oil prices by adopting other fictitious pricing systems. The first of these

5. Wayne Leeman in his book *The Price of Middle East Crude Oil*, (Cornell University Press, 1962 page 92), estimates that the abandonment of the Single Basing Point System and the adoption of the Double Basing Point System led to an actual reduction in the price of Middle East oil from $2.95 per barrel to $1.05 per barrel.

was called the 'Equalization Point'; it assumed a hypothetical competition between Middle East oil and US oil at a certain point of equalization in the consuming areas. This was initially selected in a centrally situated area (Naples, Italy). Under this system, the price of the Middle East oil exports from the Gulf plus transportation costs from the Gulf to Naples should be equivalent to the price of US oil in the Gulf of Mexico plus its transportation costs to the port of Naples. This led to a reduction in the posted price of Middle East oil by an amount equivalent to these transportation charges from the Gulf point in the Middle East to Naples. By shifting this hypothetical competitive Equalization Point westwards to London, the price of Middle East oil was once more reduced by an amount equivalent to the transportation costs from Naples to London. And, by moving this point further westwards (New York), the price of Middle East oil was reduced still further by an amount equal to the difference in transportation costs from London to New York.

Besides these unilateral price cuts on Middle East oil, further cuts were caused by differentials between Middle East and US oil price fluctuations. When the companies decided on a general price increase following oil shortages, such as those immediately after the Second World War or the Suez Crisis, the increase in price of Middle East oil was less than that of the Western hemisphere and, by the same token, when the companies decided on a general price cut, as in 1959, the cuts on Middle East oil were heavier than those applied to the Western Hemisphere. These changes in the relationship of the oil price structure between the two regions resulted in a continuing weakening of the oil price linkages between the Middle East and the US. The last unilateral price cuts of Middle East oil by the companies in the summer of 1960 marked the end of these linkages.[6]

6. According to the price comparisons of Professor Edith Penrose in her book *The Large International Firms in the Developing Countries* (op. cit. page 184), concerning the evolution of the oil price relationship of the posted price of Middle East oil as a ratio to that of a similar oil exported from the United States the price of Middle East oil became in 1961 54 per cent of that of a similar American oil (against 100 per cent in accordance with the Double Basing Point System, whereby the FOB price would be equalized).

The subsequent structural changes in the world oil market, referred to above, and particularly those pertaining to the emergence of a free market allowing for some competition among sellers and buyers, were inevitably reflected in changes in the oil price structure. This was despite the fact that the size of that market was marginal when compared with the amounts of oil entering the total international oil trade. Thus, there appeared for the first time the phenomenon of a 'realized market' price for the sale and purchase of crude, i.e. a real commodity exchange and transfer between buyers and sellers.

But these 'realized market' prices were nonetheless subject to the same price leadership by the companies, simply because the realized price took the form of a discount on the posted price, which was set by the same companies. More important was the fact that the price leadership of the major companies stemmed from the general level of the per barrel tax-paid cost incurred by the companies against lifting oil, including the taxes, royalties, and other charges paid to the host government under the terms of the concession. In the Middle East, where the companies' control was almost total, there was more or less a uniformity in the level of the tax-paid cost, because of the uniformity in the financial terms for the concession, as well as in the basis for setting the prices for various crudes. In fact, the concession system and company practice in the relationship with host governments was based on a similar financial treatment of all the producing countries in the Middle East. Almost all the concessions in that area contained a clause of 'most favoured country' financial treatment which made any advantage gained by any one producer country applicable to the other producer countries. This, together with the unified structure for the posting of all prices, secured uniformity of the tax-paid cost for oil liftings (this clause also helped the companies to resist any demands from a producing country on the grounds that other producing countries would claim the same!).

The uniform tax-paid cost was the basis on which the companies priced the oil they offered for sale to Third Party buyers. By adding such profit margins as were deemed necessary, the companies were able to generate the required cash flow for the finance of their world-wide vertically and horizontally integrated operations. It was this price (the tax-paid cost plus the profit margin) which influenced the price for other sellers, for it formed the 'floor price', below which the companies could not go any further. Cutting prices in the market below this floor would affect the companies' and their investments, since it entailed lower profits, or in some cases, losses.

In the new producing areas outside the Middle East, the tax situation was different, and as was mentioned earlier, created disparities for the lifting costs of that oil in relation to the Middle East. At a time, when the basis for calculations of the tax-paid cost for Middle East oil was uniform, being based on a constant posted price, and hence constant tax and royalty ratios, the new forms of investment relationship provided a different basis for the calculation of those taxes and royalties, being made mainly on the basis of actual sales of crude oil in a free market. In other words, prevailing prices in the market were those which determined the amount of tax payable to the host government of the new producing region. Or, to put it another way, at a time when the per barrel government-take in the Middle East was constant, that 'take' in new producing regions was fluctuating with varying 'realized market' prices of crude sales on the free market. This applied even to the companies which did not sell exported crude to a 'free market', but exported it for use in their own refineries, because tax calculations were always based on the strength of sales contracts to Third Party buyers. It was, therefore, in the interests of the independent oil companies to sell oil at low prices in order to reduce the tax payable to the host governments.[7]

Furthermore, whereas Middle East oil was generally marked by very low costs and few, mostly depreciating investments, the costs of oil production in newly opened regions were very high owing to the rise in the volume of

7. The existence of this system in Libya has resulted in a very low level of government-take. In certain cases, this level was brought down to less than 50 US cents per barrel.

the investments, the high degree of risk and the low rate of depreciation (because of the newness of oil investments). These disparities affected the profit margins of the independent investment companies. In addition, independent investors in those new areas suffered from a weak cash flow and enormous costs in borrowing the necessary capital, unlike the major oil companies, who were powerful financial entities and enjoyed in most cases, the capacity of self-financing their own investments.

All these factors induced the independents, followed by the Soviet Union,[8] to offer crude at relatively low prices, within the framework of a price structure determined by the major companies and geared to the level of Middle East tax-paid cost.

Clearly, the price cuts made by the newcomers threatened the dominant market position of the major oil companies. They were the first signs of the major oil companies' weakening position, and their reduced control of a free market now began to appear. Consequently, these new market relationships created an incentive for the major oil companies, to reduce the tax-paid costs borne by them in the Middle East, in order to preserve their competitiveness, and hence their control of world oil markets. Reducing the tax-paid cost served also to increase their profit margins for the financing of new investments undertaken by the companies in the new producing areas. This was why the American companies, followed by other major companies, undertook oil price cuts in 1959, followed by unilateral price cuts of Middle East oil prices in 1960. The oil producing countries' revenues were proportionately cut. As a direct result, representatives of the leading Middle East oil producing countries,[9]—Iraq, Iran, Saudi Arabia

8. The reasons that reactivated the Soviet Union to offer lower priced crude oil were different, for what mattered for this centrally planned economy was not the cost-price relationship, as much as the ability to procure foreign exchange from oil exports. Oil provided for the Soviet Union something like 40 per cent of its total foreign exchange earnings.
9. OPEC membership was however later expanded to include, successively, Qatar (1961), Indonesia (1962), Libya (1962), Abu Dhabi (1967), Algeria (1969), Nigeria (1971), Ecuador (1975) and Gabon (Associate Member 1973, Full Member June 1975).

and Kuwait—plus Venezuela, met in Baghdad in September 1960 at the invitation of the Iraqi government—the meeting gave birth to the Organization of Petroleum Exporting Countries (OPEC) which took the political decision to put an end to any further price cuts by the oil companies.

Price relationship in the period following the creation of OPEC

Although OPEC's successes in the first ten years of its life were rather limited and modest, it was nevertheless those successes which marked the process of change and reversal in the power relationship between the producing countries and major international companies, which ended in a series of transformations in the structure of the international oil industry.

OPEC's action on pricing during the first phase of its life can be summarized as follows:

(a) The creation of OPEC was by itself a collective political act, taken by its founding members to put an end to the oil companies' absolute freedom in setting oil prices. According to OPEC's first resolutions, the oil companies were not permitted to undertake any price cut or amendment without prior consultation with the governments of the producing countries. Timid though it was, that decision was the first event in the history of international economic relations to give a practical significance, albeit a limited one, to the concept of the inalienable right of the developing countries to sovereignty over their natural resources. By seeking member states' solidarity and their collective action to safeguard their common interest, OPEC in fact ushered in a new era of international pricing relationships, whereby the issue of oil prices became part of state sovereignty. Thus, the establishment of OPEC as an instrument of collective action by the producers can be regarded as the first turning point in the international economic relations towards the states' control over natural resources. More important, that event can be considered as the first amendment—through the application of the principle of sovereignty—of the oil concession system itself. Prior to the establishment of OPEC, governments of the producing countries were unable individually to stop the companies from exercising what they, the companies, considered to be one of their rights, namely to decide on the level of prices in accordance with the terms of the concession.

Therefore, despite the failure of OPEC in its early years to apply its resolutions on restoring oil prices to their pre-1960 levels, the collective decision to freeze oil prices was a basic change in government's attitudes vis-à-vis the oil industry. For the first time, the producing governments entered as a partner in setting prices, although a sleeping one. Consequently, this development was to be an initial step in a long process of change in the government-companies relationship, whereby collective action was found to be the only effective means of restoring the right of sovereignty and changing the status quo of the international economic order, of which the raw material producers/exporters are the real victims.

(b) Although OPEC failed during that period to put into force many other decisions, it nevertheless established new trends and new concepts in the area of oil pricing. In particular, OPEC emphasized the exhaustible nature of oil wealth as the only valid parameter that should govern long-term pricing policies. It also drew attention to the long-term relationship between the oil price and the problem of economic and social development requirements in producing countries. Furthermore, it brought to the surface the concept of fair prices, besides emphasizing the compelling necessity of improving the terms of trade for oil and preventing the erosion of the oil purchasing power caused by world inflation. Moreover, OPEC called, in more than one resolution, for a production regulation as a basis for more rational pricing, and so on. Deprived of their teeth though they were, these decisions constituted, nevertheless, a solid basis for the measures which OPEC later adopted in its stand against the companies, successfully applying many of the concepts which, prior to the emergence of OPEC, were alien to international relations.

(c) The decision to stop the major oil companies from reducing prices led in practice to breaking a vicious circle of price cuts which, it was estimated, would otherwise have occurred. The emergence of a free market for oil, and the entry of new investors into the oil industry in search of a share in that market (through giving discounts on prices), as well as the existence of a new formula for investment and taxation disturbed the old company-dominated market equilibria. One major result of those developments was the disparity in calculating the per barrel tax-paid cost incurred by the large old investors versus that of the small newcomers. This always created the incentives for the older companies to undertake numerous reductions in the tax-paid cost, in order to preserve their leading and central position in the pricing system and their share of the market. Under the new market conditions, cutting posted prices was the most effective means to restore the price leadership of the concession-holding companies in the Middle East. The reason was that the per barrel net government-take, and hence the tax-paid cost in the Middle East was computed on the basis of the posted price, while that in the new producing areas, paid by the newcomers, was computed on the 'realized market' price. It was a feature of this disparity in the calculation of the tax-paid cost that it created a permanent tendency on the part of the newcomers to depress the 'realized market' price as the only means of reducing their tax burden. Such a reduction on the part of the newcomers forced the old investors to cut the posted price as a means of reducing the fiscal cost on them, and of maintaining their competitive position. The problem however was that any reduction in the posted price would in its turn lead to a further price depression in the market, because of the fact that the 'realized market' price took the form of discounts on the posted price, and that the tax-paid cost borne by the large companies always formed the floor for the 'realized market' prices. Therefore, cutting the posted price could lead the newcomers into further cuts in the 'realized market' price, and so on.

Thus, the decision of OPEC to freeze the posted price led actually to the creation of an almost constant price relationship between the tax-paid cost of Middle East oil and the realized price in world markets. This meant that there was no longer any financial advantage to the major companies in granting large discounts on the posted price of the oil placed by them on the free market, except only to the extent of realizing sufficient profits for the expansion of their operations under new market conditions where the independent companies were no longer in a position to increase their market share.

(d) Such a potential race in price-cuts was more or less completely halted when OPEC worked successfully to remove this disparity in the method of calculating the per barrel tax-paid cost that existed between the old and new regions. In extending its membership to include new producing countries which adopted different tax structures, a standardized fiscal relationship with the investors, based on fixed prices, was finally reached. The new members undertook such legislative measures as to bring those relationships into parity with the traditional concessionary tax pattern. Libya, for example, in 1965 had amended its Petroleum Law, to bring it into parity with that of the rest of the Middle East, i.e. taking the posted prices as a basis in calculating the tax-paid cost, instead of the previously held fiscal regime, whereby the companies were allowed to enjoy a lesser per barrel tax-paid cost computed on discounted realized market prices. In other producing countries adopting fiscal regimes other than that of profit-sharing based on the posted price, like Algeria, agreements were reached between the host governments and the operating companies for the determination of a fixed price level to serve as a basis for taxation (the tax-reference price). As in the case of posted prices, this latter price did not allow for any deterioration in the per barrel net-government take.

The result of these OPEC price disciplines was that a more or less uniform, fixed and unvaried level of tax-paid cost was applied in almost all producing countries. Under the new fiscal conditions successive cuts in the realized market price for oil were no longer made, since no advantage accrued to investors from cutting the realized market price, once the per barrel government-take (i.e. the cost to the companies) came to be decided on the posted price, which was held constant by OPEC decisions, irrespective of the realized market price movements. The unification of the tax regimes thus created the incentive for the major oil companies to stabilize the price for oil offered by them to all purchasers, at a level descending below which would mean a loss to them and would affect their world-wide investment operations. This was in contrast to the former situation where, because of the disparities in the fiscal structures among various regions and different operations and investors, the oil companies had a permanent interest in transacting business with Third Party buyers at lower prices, so as to create a lowered base for calculating the government-take, a situation, which if continued, would have led to a race in cutting prices between major oil companies working in the Middle East and the newcomers operating in the new producing countries. This is why the period between 1965 and 1969 witnessed a relative stability in the market price, albeit at a low level.

(e) The decision on royalty expensing taken by OPEC, despite its partial application, played a role in strengthening the oil price structure, in so far as it raised the tax-paid cost, and hence the prevailing floor prices, because applying this formula, even in a limited way, led to an increase in the government-take per barrel. The royalty expensing formula was based on the principle of separating the royalty, (traditionally rated at 12.5 per cent of the posted price), from the tax, amounting to 50 per cent of the posted price against the previous regime which considered the royalty as part of the tax liability. Such a separation meant that the royalty was payable to the government irrespective of the size of the taxable profit, provided that royalty payments were to be deducted from those profits. In other words, the royalty was to be paid separately from the 50 per cent, but to be considered at the same time as an expense or cost in computing taxable profits. This concept brought about a net improvement in computing the government share, as its full implementation would lead to increasing the net per barrel government-take from 50 per cent of posted prices to 56.25 per cent.[10] But when that principle was implemented according to agreements concluded separately between the governments and companies in 1965, OPEC countries failed to fully enjoy that increase, as it allowed the companies discounts on the posted price for the purpose of computing the royalty, but in a decelerating manner, so as to phase out discounts through 1972. Those discounts meant in practice a modest net increase in the government-take for the first year of the implementation of the agreement that did not exceed 3.5–5.5 US cents per barrel, compared with 11 cents per barrel had the royalty been fully expensed immediately.

10. Prior to this amendment, the government share was half the difference between the posted price 'P' and production cost 'C', thus the government-take = 0.5 (P–C). With full expensing royalty the government-take would be as follows:

$$0.125P + 0.50\,[P - (0.125P + C)] = 0.5625P$$

However the important factor was not the size of the increase in government revenue[11] as much as the creation of an upward trend in the level of the tax-paid cost, and hence a strengthened 'realized market' price, a trend which was later reflected in far more radical measures by OPEC.

The first significant change in the pricing power relationship: the Tehran and related agreements

As the 1960s closed, a new turning point occurred in the OPEC-companies relationships towards a greater self-assertion by the oil producing countries over the control of their oil resources. The initial impetus came from a member country, Libya, where the success of the revolutionary government in solving its outstanding problems with the companies, mainly on the issue of pricing Libyan oil, opened a new page in the history of those relationships. Libya, as is known, had old claims on the companies concerning various oil issues, most important of which were the protests annually lodged against the companies for pricing the light low-sulphur and short-haul Libyan crude at a level similar to those fixed by the companies for heavier and higher sulphur crudes in the East Mediterranean terminals. Following the failure of successive negotiations with the companies, a real show-down took place in 1970, when the government issued instructions to substantially cut back (up to 30 per cent) production rates of many independent companies working in Libya. Conditions were in fact propitious in that country where, unlike the Middle East where a monopoly position was held by the multi-nationals, over 20 companies were operating inde-

11. See Zuhayr Mikdashi: *The Community of Oil Exporting Countries* (Cornell University Press, 1972). The principle of royalty expensing was accepted in 1963, and in accordance with the agreements reached in 1964 with the companies, those discounts started by 8.5 per cent in 1965 and decreased annually to vanish completely eight years later. Thus in the year when the formula was effectively implemented (that is 1965), the discount was 7.5 per cent, the net government-take for Arabian Light had increased from 85 US cents per barrel (on a posted price of US $1.80 per barrel and assumed production cost of 10 US cents) to about 89 US cents per barrel (against a government-take of about 97 cents in case of full royalty expensing). This amount was decreased by further small discounts granted on gravity differentials with a resulting net increase in the government-take of less than 4 US cents per barrel.

pendently from each other. This allowed the revolutionary Libyan government to efficiently put enough pressure on the companies separately, especially those who had no access to any other crude oil elsewhere, such as Occidental. The strength of its financial position, through past accumulated reserves, added a further bargaining power to the Libyan government in restoring its rights. As a result, the companies accepted, one after the other, the Libyan demand of correcting the price upward by 30 US cents per barrel. Furthermore, Libya obtained an upward correction of 5 per cent of the tax ratio (from 50 per cent to 55 per cent) in lieu of the accumulated financial claims resulting from previous government protests. In addition, Libya succeeded in introducing, for the first time, the principle of annual price increases, albeit modestly (2 cents per barrel through 1975).

The Libyan production cut-backs happened at a time when demand for OPEC oil was growing at faster rates, reflecting the higher economic growth of industrialized countries. Demand pressures were added by the entry of the United States of America into world markets as an important buyer of increasing quantities of OPEC crude. A further element of strain on the supply of short-haul crudes was the political occurrence of the explosion of the pipeline transporting part of Saudi production (about half a million barrels per day) to the East Mediterranean terminal of Sidon. These short-haul crude oil shortages resulted in a sudden reversal in market price movements. For the first time 'realized market' prices for all short-haul crudes (Iraqi and Saudi crudes from the East Mediterranean terminals, Libyan, Algerian and Nigerian crudes) went substantially beyond posted prices.

Under such conditions, OPEC began adopting an entirely new approach in dealing with the problem of prices and negotiations with the companies.[12] After a long phase of

12. It should be emphasized that the shortages of the short-haul crude were not the real cause for this shift in government-companies relationships, as such shortages had occurred once in the past without resulting in any change in the price system. Following the closure of the Suez Canal during the June 1967 Middle East War and the resulting increases in the freight rates, OPEC convened a meeting of the Conference in Rome to deal with the issue of the oil

sleeping partnership, during which its sole action was to prevent the companies from undertaking any price cuts without prior consultation with the governments of the oil producing countries, OPEC started to emerge as a more active partner in determining the price of its oil. Thus, at the 21st Meeting of the OPEC Conference in Caracas in December 1970, and in accordance with its Resolution 120, OPEC decided to enter into collective negotiations with the companies on the strength of specific demands that were laid down in the Resolution; namely, the tax ratios were to be amended upward in line with the Libyan settlement, posted prices were to increase substantially in reflection of the recent market increases, all discounts and price rebates enjoyed by the companies, including those of royalty expensing, were to be completely deleted, and the system of price differentials among the various degrees of API to be revised, etc.

Collective negotiations were to be conducted on a regional basis, and not OPEC-wide, by forming a negotiating ministerial committee composed of the oil ministers of Iraq, Iran and Saudi Arabia jointly to negotiate with the companies on behalf of the OPEC member countries bordering the Gulf (Kuwait, Qatar and Abu Dhabi, besides the other three Gulf countries mentioned above). After more than

price in reflection of the new situation. However, OPEC failed to change the price relationship, and finally accepted a formula that was individually signed with the companies by the short-haul crude oil producing countries, (Libya and Saudi Arabia). According to this formula, the companies compensated those producers by conceding the right to discounts for the purpose of expensing royalty (which meant that for those crudes, unlike the Middle East crudes, the royalty was fully expensed), and resulted in a net increase of about 7–8 US cents per barrel. This small amount was, of course, much less than the market increases. More important was that the formula maintained the OPEC price structure as defined by the companies, and isolated the short-haul producers from the long-haul producers of the Gulf. (Those extra cents were also applied to Iraq in spite of the fact that the royalty expensing agreement was not signed with the Iraqi government, because of the conflict between the government and the companies on Law No. 80). By accepting such a meagre settlement, OPEC had actually lost a very good chance of reversing the government-companies pricing relationship, because of the strong bargaining power that it was actually enjoying as a result of the market shortages. But the readiness of the member countries to enter individually into agreements with the companies enormously reduced the OPEC countries' bargaining power.

one month of difficult and interrupted negotiations, the Tehran Agreement was successfully concluded with the oil companies in mid-February 1971. This agreement, together with other related agreements subsequently concluded,[13] created a new pricing system that was intended to last 5 years (through 1975), but which collapsed more than two years later under the pressure of oil events. However, the Tehran negotiations represented a unique experience of its kind, whereby OPEC resorted for the first time in its history to joint action and collective confrontation with the companies as the only means towards fulfilling its legitimate demands. This turning point in the government-companies relationship was, in fact, the first development to indicate a real shift, albeit limited, in the oil power towards a more effective exercise of the state's right of sovereignty over its natural resources. By evoking the principle of collective legislation to enforce its demands on prices, OPEC had already changed the power relationship partially in its favour, by shaking up the old system of the companies' price domination. This happened in the historic Extraordinary Meeting of the Conference held in Tehran prior to the conclusion of the agreement, whereby the heads of OPEC delegations declared solemnly their governments' readiness to enforce OPEC's demands on the prices by legislation, if necessary, should negotiations with the oil companies fail. They also declared their governments' readiness to impose an oil embargo on any operating oil company that did not abide by such legislation.

The oil companies, in response, resorted—with the backing of their governments—to new approaches in strengthening their bargaining position. Hence the waiver, con-

13. These were the Tripoli Agreement, concluded in April 1971, for the Libyan oil, the Lagos Agreement for the Nigerian oil, the East Mediterranean Agreement, concluded with Iraq, concerning Kirkuk oil exported from Tripoli/Banian, and the agreement concerning the Saudi oil exported from Sidon. The Tehran pricing system included also currency protection agreements, concerning the price adjustments to offset variations in the exchange rates of the US dollar (the currency denominating the price of oil as well as the currency of oil payments) vis-à-vis the major world currencies. Geneva I Agreement was concluded in January 1972, following the first devaluation of the US dollar, and Geneva II Agreement was concluded in June 1973, following the second devaluation of the US dollar.

sented to by the US government, of the Sherman Anti-Trust Law, allowing the companies collectively to enter into negotiations with the OPEC countries, and also the inter-company arrangements for redistributing crude oil among themselves in case of any one of them suffering from crude shortages. This was besides the various political pressures applied by certain governments of the oil consuming countries, to contain the new development.

The pricing system of the Tehran Agreement

The main features of the Tehran pricing system were the following:

(a) A general across-the-board increase of 35 US cents per barrel in the price of all crudes of the Gulf area;
(b) An annual increase of 2.5 per cent in lieu of the erosion of the purchasing power of oil revenues, caused by world inflation, together with 5 US cents per barrel in lieu of any increases in the price of products in the major consuming areas. (The latter increase was not spelt out clearly in the agreement but was part of an understanding among the negotiators.)
(c) An increase of 5 per cent in the tax ratio (which was already agreed and applied by the companies in the case of Libya, Iran, etc.), and the result of which was to increase the per barrel government-take from 56.25 per cent of the posted price to 60.625 per cent.[14]
(d) All the discounts on the posted price enjoyed by the companies were abolished. These included the marketing discount of half a cent per barrel (after being successively reduced from a much higher discount rate previously enjoyed by the companies, originally 2 per cent of the posted price), and the discounts on account of royalty expensing.[15]

14. The per barrel net government-take:
$$= .125P + .55\ (P - .125P - C)$$
$$= .125P + .55\ (.875P - C)$$
$$= .60625P - .55C$$
It should be recalled however that the previous ratio of 56.25 per cent of the posted price was not fully implemented at the time when the Tehran Agreement was signed, because of the decelerating discounts enjoyed by the companies, as was explained earlier.

15. The abolition of the discounts led to the full implementation of the principle of royalty expensing, i.e. a further increase in the government-take by about 3–4 cents per barrel; (the scheduled discount on royalty expensing in 1971 was 2 per cent).

(e) The system for the price differentials of various OPEC crude oils (the evaluation of each full degree of API as a gross indicator of the quality-yield differentials) was revised. According to the revision, the linear relationship of 2 US cents per barrel per each full API degree was reduced to 1½ US cents.[16] Furthermore, the new system, as opposed to the old one, gave a value for the fraction (decimals) of the API degree. This re-evaluation of the price differentials led in practice to the upgrading of the heavy and medium crudes, including the Arabian Light, due to the fact that the application of the general price increase was on the Qatari Dhukan, 40 API, the result of which was an increase of the prices of most of the Gulf crudes, reaching as much as 5 US cents per barrel for heavier crudes.

All these provisions had led to a general increase in the tax-paid cost of lifted oil from the Gulf area of about 45 US cents per barrel in 1971.[17] A similar increase was, of course, reflected in the realized market price, thus substantially strengthening the OPEC market price structure.

These changes in the new pricing system were not confined to the provisions of the Tehran Agreement, as new concepts were introduced in subsequent related agreements. The agreements, such as those dealing with short-haul oil, like that concluded with Libya, admitted for the first time into the price structure the element of quality differentials other than the yield differential, as grossly measured by API: namely, the sulphur content of crude oil. The agreement attributed a value for the sulphur content of the low-sulphur lighter African crudes that escalated over time. The previous pricing system had completely ignored this quality differential, at a time when environmental restrictions in the main markets of OPEC crude, i.e. Western Europe and Japan, were not so severe as to create a

16. This linear relationship applied to crude between 30 and 40 API. Those which were under 30 and over 40 were to be dealt with separately by each member state, in agreement with the operating oil company.
17. This amount includes the scheduled price increase of 1972, which was brought forward to be applied as from 1 June 1971, as a compensation for the OPEC countries' demand to implement the Agreement retroactively to 1 January of that year (the effective date of the Agreement was 15 February 1971).

special demand for the low-sulphur light crude. The US, where such restrictions already existed, was not up to that time an important buyer of OPEC crude. By the beginning of this decade the pattern of demand for crude oil changed both in terms of the yield pattern, and the growth of environmental restrictions in Europe and Japan, at a time when American demand for lighter, lower sulphur OPEC oil began to visibly increase. Instead of a fuel oil-based energy structure in Europe, a shift towards a market pattern based on the lighter ends of the barrel started to assert itself, reflecting a trend that began to coincide more with the US energy consumption patterns. Likewise, environmental restrictions were growingly imposed on refiners who, as a result, were obliged to blend increasing quantities of sulphur-free crude with sulphurous crudes from the Middle East. The alternative for them was to undertake investments in cracking and hydro-reforming and desulphurization, which would be inevitably reflected in a higher value for low-sulphur (and lighter oil as will be discussed later on) crudes in the market. Hence the Tripoli Agreement provided for a low-sulphur premium to be added on the price, over and above the general price increase decided by the Tehran Agreement. Initially, that premium was 10 US cents per barrel (for crude oil with less than 0.5 per cent sulphur), to be escalated to reach 20 US cents per barrel by the end of 1975.

On the other hand, agreements on short-haul oil embodied another new concept concerning the geographical location premia and the variation in freight rates. The aim was to syphon back the additional windfall profits reaped by the companies as a result of tanker shortages and higher freight rates resulting from flare-ups in the tanker market. The previous pricing system did not take account of such variations in the geographical location premia, since it based the price structure on a fixed freight relationship between the short-haul oil and long-haul oil, irrespective of the variations in the tanker market. Thus, under the old system, prices of the short-haul oil were hardly sensitive to market variations. In opposition to the old pricing regime, the Tripoli Agreement of April 1971 introduced the concept of a moving freight scale to serve as a basis for adding or removing any additional premium for the price of short-haul crude that might occur as a result of tanker shortages or increases in freight rates. Accordingly, the Agreement admitted a special premium of 13 US cents per barrel against the closure of the Suez Canal, a premium to be removed by stages once the Canal was re-opened. Furthermore, the Agreement admitted another 13 US cents per barrel as a moving premium to be added to the price in compensation for the rise in freight rates. That amount was to vary upward and downward in the light of the movement in the freight rates as measured by the World Scale. It was to be added in full whenever the World Scale reached 92 points (above which no more premia were to be added to the price) and to be totally removed when the World Scale went down to 72 points (below which no price reduction was to be made). Within this range freight premia were to move by about half a cent per barrel for each full point variation in the World Scale.

The Tehran pricing system and the problem of the US dollar

More important in the new pricing regime was the acceptance by the oil companies in accordance with the subsequent Geneva I and Geneva II Agreements (which made part of the Tehran Pricing System), of the principle of protecting the price of oil against the variations in the exchange rates of the US dollar, against the other major world currencies. Being the unit of account for denominating the value of oil, (as well as the unit of payments), the US dollar has far-reaching impact on the real purchasing power of the oil revenues in terms of other currencies. A depreciated dollar would bring less currency than an appreciated one, and hence a lesser purchasing power in that currency. By the end of the sixties the international monetary system was already beginning to change in a manner that would allow currency fluctuations to affect the stability of the oil purchasing power. During the post-war period, the monetary system that evolved with the Bretton Woods system provided enough protection for the oil revenues, partly because of the fixed parity system of money exchange, and partly because of the strength of the dollar itself at a time when the reconstruction of Europe and Japan was suffering from a dollar shortage. The first signs of a crack in this system began to appear with the shift of the US financial status from a surplus country to a deficit country, as was reflected in the movements in the US balance of payments. This led to mounting pressure on the dollar that seriously threatened the whole Bretton Woods system, and hence to a further weakening of the dollar. This in turn meant that oil revenues were adversely affected as far as purchasing power in terms of other currencies were concerned. Such adverse effects of changes in the exchange rate of the dollar against the other major currencies would, of course, depend on the pattern of trade of oil exporting countries, in the sense that such effects would increase as the trade of OPEC with the US decreased in favour of trade with other major industrialized countries. At the OPEC Conference in Caracas, a resolution was passed outlining measures that should be taken to protect the price of OPEC oil against an eventual depreciation of the US dollar. The monetary situation, however, further deteriorated when the US decided to stop the convertibility of the US dollar against gold and the consequential monetary disruptions that were reflected in the devaluation of the US dollar and the floating of other major currencies in the light of the money market movements. Accordingly, negotiations with the oil companies were collectively conducted on behalf of the OPEC countries bordering the Gulf, that ended with the signing of an agreement in Geneva in January 1972, whereby the price of oil was corrected upward by 8.5 per cent to compensate fully for the loss in the purchasing power of the oil revenues resulting from this effective devaluation of the dollar. Furthermore, the agreement introduced a mechanism of quarterly price adjustments to take care of future market variations in the exchange rates of the US dollar against the other major currencies. The measurement of change in the value of the US dollar was based on the movement of the exchange rates of that currency against 9 international major currencies, and in relation to the IMF parities as they stood on 30 April 1971, the date preceding the US decision on the non-convertibility to gold of the US dollar. Therefore, whenever there was a variation in the average exchange rates of those currencies in any quarter against the dollar, that exceeded two full points, up or down, from the starting average, (i.e. the average of the exchange rates of those currencies as they stood on the day of the signature of the agreement, which was 11.02), the oil price would be corrected proportionately either upward, to offset the loss in the purchasing power resulting from a depreciating dollar, or downward, whenever the dollar appreciated in the market (downward corrections can never descend below the floor of the Tehran price structure).[18] The concept of that mechanism was that oil price movements should be such as to prevent any deterioration in the purchasing power of the oil revenue in terms of other currencies, but also not to increase that purchasing power beyond the level that was determined in accordance with the Tehran Agreement, i.e. preserving the same purchasing power of the Tehran prices in terms of other currencies.

However, the implementation of that agreement showed certain shortcomings in the mechanics of compensation, so that in many cases oil revenues were not fully protected. Consequently, and following a second effective devaluation of the US dollar, a revision of the Geneva Agreement, called Geneva II and concluded on 1 June 1973, was made, again through collective negotiations. In accordance with

18. The basket of currencies for the measurement of the variations of the exchange rate of the dollar was composed of the pound sterling, the Swiss franc, the French franc, the Belgian franc, the Deutschmark, the Japanese yen, the Dutch guilder, the Italian lira and the Swedish kroner. As to the mechanics of future changes, the following equation was adopted and was based on the market quotations to be regularly provided by the National Westminster Bank Limited, London:

$$\text{The adjusted price: } = 0.849 \left(\frac{T \times B - A}{11.02} \right)$$

whereby T is the posted price as defined by the Tehran Agreement, A is the most recent average change and B the average change in the preceding period. The amount of 11.02 represents the starting average of the exchange rate cumulative changes of these currencies against the dollar since 30 April 1971 until the day of the signing of the Agreement, (the factor .849 represents the initial price correction to offset the official devaluation of the dollar).

the revised agreement, a new mechanism for a monthly measurement of change actually provided for a more adequate protection of the oil revenues.[19]

The significance of the change in the pricing system

The new pricing order, based on Tehran and related agreements, brought the pricing relationships into a phase in which the oil producing countries played a more assertive role in the operations of the international oil industry:

(a) The system brought OPEC forward for the first time as a world power that was formally recognized, not only by the oil companies, but also by the large energy establishments in the industrialized countries, if not by governments. Prior to this development the oil companies had ignored OPEC as a collective bargaining power and insisted, as was the case in the negotiations for the royalty expensing, on conducting separate negotiations with governments of the oil producing countries, to keep their negotiating power weak. The Tehran Agreement contained, for example, explicit reference to many OPEC decisions, mainly those taken in the Caracas Conference.

(b) The new system provided for a negotiated price structure destined to last for some time (through 1975), after which the price was to be re-negotiated in the light of ensuing developments. This feature is an explicit confirmation of the administered nature of the oil prices, irrespective of the market variations. It was also a clear indication to the possibility of price planning that could be conducted by the producers, for a certain period of time, irrespective of the relative movements of supply and demand. So, the scheduled price escalations on account of inflation and other factors, as well as the price adjustments in the light of exchange rate changes, were all reflected in the contracts of

19. The basket of currencies was enlarged to include both the Australian and Canadian dollar and applying the following equations:

$$\text{The adjusted posted price } (P'): = \frac{P + T'(B - A')}{100}$$

whereby P is the posted price that would have applied on the first day of the month of adjustment (absent such adjustment), T′ is the posted price that would have applied under the Tehran Agreement, excluding the effect of any currency adjustment, A′ is the most recent average change of the currencies against the dollar and B′ is the new average.

oil purchases. In those contracts, price escalations were generally provided to the increase in the tax-paid cost in accordance with the new escalated price structure. It is by the new system, however, that price administration was for the first time to be jointly conducted with OPEC which emerged as a new, albeit still timid, controlling power in the international oil relations.

(c) The new regime has clearly shown that changing the status quo of international economic relations towards more equitable sharing of the oil wealth between producers and consumers, could only be made through powerful and more active partnership on the part of the producing countries. This feature was to play later on a greater role in changing more substantially the pricing relationship.

The pricing system under the Agreements on Participation

The arrangements for participation did not have direct effects on the posted-price system, but their effects on the free market pricing system were dramatic. Among the most important features of state participation in oil operations was the acquisition by the government of a share in the production for itself, to dispose of at prevailing market prices. This meant that the revenue which derived from the sale of the government's share equalled the full market price, whilst the revenue deriving from lifting the companies' share equalled the tax-paid cost on the companies, which itself was less, by a certain proportion, than the posted price, and invariably less than the market price. The original idea was that resale of the government's share to the companies would be carried out at the market price, and this is what the participation arrangements actually produced, although at the start things were different, in that resale took place at prices below market price.[20]

The lifting by the companies of part of the oil at the market price against their own share at tax-paid cost led naturally to an increase in the average per barrel cost incur-

20. At the beginning the arrangements for reselling the government's share (Bridging oil and Phase-in oil) enabled the companies to enjoy larger discounts, since they took this share at tax-paid cost plus a fixed amount and/or at the Quarter-way price plus (see above).

red by them. Clearly, the greater the companies' lifting of government share, the higher the prices of oil placed by them in the market, simply because the rise in the cost should be reflected by a proportionate rise in the 'floor' price. As the average cost increased, because of the participation and the increasing share of the government to be bought-back at higher cost than the equity on tax-paid cost, so the difference between the prevailing market price and the cost borne by the companies narrowed, i.e. their profits dropped and, in consequence, the structure of the price strengthened.

Cost increases of this sort reduced the ability of the companies to lower market prices and their leeway to grant discounts to Third Party buyers. Naturally, circumscription of the companies' freedom of manoeuvre in the market depended on the cost structure for their lifting of oil, and this in turn depended on two basic factors. The first was the share of government participation, for, as this share rose, so the cost increased owing to the buy-back by the companies of that share at a price higher than the tax-paid cost. Therefore, when the government's share was 25 per cent, the cost to the companies was low, because the greater part—75 per cent—was taken by them at the tax-paid cost, which was much lower than the market prices at which the smaller share (25 per cent) was bought-back from the government. Therefore, the expansion of the government's share to 60 per cent altered the picture by a considerable increase in the cost to the companies. The second factor concerned the level of tax and royalties imposed on oil lifted via the companies' share and their ratios to the posted price. As those ratios were low, so the cost on the companies was reduced, and this was true in reverse, i.e. the increase in the tax ratios meant a proportionate increase in the cost. Therefore, when the tax was 55 per cent and the royalties 12.5 per cent, or approximately 60 per cent of that posted price, the profits of the companies from their share were high indeed. As a result, their capacity to control the market price was broad. But by raising the tax ratio to 85 per cent and the royalty to 20 per cent, or 88 per cent of the posted price, the cost to the companies increased to an extent which greatly reduced their liberty of movement in the market and their ability to allow discounts on prices, because of the reduction in their per barrel profit margins. The fusion of these two factors played a major part in raising the price levels of sales to Third Party buyers, because the companies' profit margin dwindled to 2 per cent of the posted price, as explained earlier. This meant that the lowest limit at which the companies could sell the oil—or the 'floor' price—was lifted considerably to a level at which it was not possible for the companies to give discounts except within very narrow limits. This ultimately signified a strengthening of the free market oil price structure.

The system of price administration by OPEC

The joint OPEC/companies price administration in accordance with the Tehran and related agreements, proved to have inherent weaknesses that made the system incapable of coping with new market developments. By the middle of 1973 realized market prices had already risen to such heights as to surpass the posted price at that time, which was on average $3 per barrel, whereas in accordance with the Tehran fiscal regime, the per barrel net government take was less than two-thirds of the price (less than $2 per barrel). Hence the enormous size of the companies' windfall profits that led the governments of the oil producing countries, partners of the Tehran Agreement, to demand the reopening of the negotiations of the Agreement in the light of changing circumstances. Negotiations in the fall of 1973 in Vienna soon proved abortive, when the companies refused the governments' demands for the adjustment of the posted price to a level that would generate a net government-take per barrel to reflect the market conditions, and to syphon back the extra windfall profits of the companies. It was only a posted price of about $6 per barrel that would generate such a net government-take. In the Vienna preliminary session of the negotiations the companies proved to be intransigent and broke off negotiations on the pretext that such a price increase would need wider political consultations, because of its very important impact on the economic activities of the industrial countries. On the refusal of the companies to continue negotiations, OPEC member countries bordering the Gulf decided to meet in Kuwait on 16 October 1973, where they announced their historic decision on the pricing of OPEC oil by the governments, independently of the companies, and in application of the principle of the right of a state to sovereignty over its natural resources. In that meeting posted prices were increased by 70 per cent, amounting to $5.11 per barrel for Arabian Light. That increase was decided in the light of effective market realizations by some of the member countries' national oil companies, who were already in the markets, and whose crude sales showed that the market price of OPEC oil was well above the posted price (exceeding in fact $3.5 per barrel).

On the following day the Arab oil exporting countries met also in Kuwait in the headquarters of the Organization of Arab Petroleum Exporting Countries, to take the political decision to impose the oil embargo, following the military and political support given by the United States of America to Israel in the October War, which was still going on at that stage. The oil shortages in the market which resulted from the succession of cuts in Arab oil production (initially 10 per cent)[21] created an entirely new situation in the market during November and December of that year, whereby realized prices were in certain cases as much as three times the posted price.[22]

Amidst those market conditions, the OPEC Conference was held in Tehran in December 1973 whereby an increase of 140 per cent was decided, bringing the posted price of Arabian Light up to $10.84 per barrel. It is interesting to recall, however, that the rationale behind the new price was not the market situation, as much as the decision to

21. Iraq took another stand in dealing with the issue, by deciding that the 'oil weapon' should be used only against those countries which supported Israel by nationalizing their interests and by imposing selected embargos. This is why Iraq nationalized American and Dutch interests in the Basrah Petroleum concession, as was mentioned earlier.
22. The National Iranian Oil Company offered 500,000 barrels a day for six months for sale through auction. Prices as high as $17.34 per barrel for Iranian Light crudes were offered by independent buyers.

relate the government-take to a level near the cost of available sources of energy. An amount of $7 per barrel was thought to represent, at the then prevailing conditions, an indication of the cost of alternative sources of energy to oil, including coal and nuclear energy. Accordingly, the level of the new posted price was computed on the basis of the fiscal relationships that were already existing, so as to produce a net government-take of that amount, $7 per barrel.

Since that important turning point in the history of pricing oil, OPEC has been the sole price administrator, in accordance with the unanimous decisions taken politically by the oil Ministers of the member states of OPEC.

However, after the initial price 'revolution' of 1973/74, OPEC in later decisions on price came increasingly to adopt a more conservative approach, whereby only the following price increases were decided:

(a) In accordance with the decision taken in Vienna in September 1975, (21 months after the Tehran decisions), the price of oil was increased by only 10 per cent, bringing the OPEC price up to $11.46.

(b) Fifteen months later a split decision on the oil price was taken in Doha in December 1976, whereby nine member countries decided to increase the price by 10 per cent to be followed by an additional 5 per cent increase six months later, and two member countries, Saudi Arabia and the United Arab Emirates did not go along with the majority decision and decided to undertake an increase of only 5 per cent. However, because of the difficulties arising from this two-tier pricing system, a re-unification of the price structure was reached during the OPEC Conference in Stockholm, in July 1977, where Saudi Arabia and the United Arab Emirates agreed to raise their prices by another 5 per cent against the other countries' agreement to give up the additional 5 per cent increase. In accordance with this price adjustment, the Arabian Light marker crude became $12.70.

(c) Two years later, in Abu Dhabi, OPEC decided to undertake quarterly price increases, the result of which was to obtain an average increase of 10 per cent throughout 1979. Thus the price for the first quarter of 1979 for Arabian Light marker crude was to be $13.335, a 5 per cent increase; for the second quarter $13.843, a 3.809 per cent increase; for the third quarter, $14.161, a 2.294 per cent increase; and for the last quarter, $14.542, a 2.691 per cent increase.

(d) However, the new situation created by the revolution in Iran and the shut-down of Iranian oilfields created mounting pressures on the price structure as set by OPEC in Abu Dhabi. Spot market prices started rocketing to unheard-of levels, to about 23 dollars per barrel in late February 1979. The OPEC price structure was unable to hold up against such market pressures, and so a new element in the price was introduced. For the first time an additional 'surcharge' was imposed by certain governments of member countries on the price of the oil exports, a surcharge which was meant to be temporary and removed once market conditions became normal.

In the face of continued mounting pressure in the market, due to the supply shortages, OPEC member countries held a consultative meeting in Geneva in late March, 1979, which was converted into an Extraordinary Meeting of the Conference, in which it was agreed to amend the Abu Dhabi decision. The OPEC price that was scheduled to be applied in the fourth quarter of 1979, was brought forward to be applicable in April 1979. This shift in applying the quarterly increases resulted in an average price increase for the whole year of 12.5 per cent instead of 10 per cent. This price adjustment was unanimously agreed on by the Conference, leaving, however, each member country free to impose such surcharges and premia as would be warranted by the market. The legality of such a decision was based on the original concept of pricing OPEC oil in the sense that OPEC pricing decisions were only minimum floor prices, below which no member country was allowed to sell, but above which every country was free to get as much as the market would allow.

(e) Meanwhile, even greater market strains were witnessed, and were reflected by a huge gap between the OPEC price and spot sale prices, which in June reached about $40 a barrel against an OPEC official price of $14.55 per barrel for Arabian Light. The escalation in the market conditions[23] thus made the OPEC price of marker crude completely unrealistic.[24] It was, therefore, decided in the Ordinary

23. Kuwait's initial premium of $1.20 increased to $1.70 and then to $2 per barrel.
24. Algeria increased its price from about $18 to $23 per barrel assuming a 'de facto' price for Arabian Light of $18 per barrel, instead of $14.55 per barrel,

Conference, held in Geneva in June, to correct the price of the marker to $18 per barrel, besides a surcharge or market premium not exceeding $2 per barrel that could be voluntarily imposed by member countries in the light of market conditions, (Saudi Arabia did not impose that surcharge). What was a new development in the OPEC price structure was the OPEC decision to impose a ceiling of $23.50 that member countries should not exceed.

(f) This structure was successfully maintained during the third quarter of 1979. It did not, however, resist the new upward pressures in the market which were strongly felt in the beginning of the fourth quarter. Consequently, prices of most of the crudes moved upwards, and those of African low sulphur crudes increased beyond the OPEC ceiling. Meanwhile, the price of Arabian Light was kept at US$18 per barrel. These successive price increases decided individually by the OPEC Member Countries have seriously affected the structure of price relationships. In fact, never before did crude oil price differentials reflect as much heterogeneity as they did during that period. For example, the price differential between Arabian Light and Iranian Light became US$5.50 per barrel in November 1979 against a historical differential of less than 15 US cents per barrel. On the other hand, the price of Sahara Blend (Algerian) became US$8.27 per barrel higher than the price of Arabian Light, whereas in January 1979 the differential in its favour has been less than US$1.50 per barrel.

(g) By mid-December 1979, and prior to the OPEC Conference held in Caracas/Venezuela later that month, the Kingdom of Saudi Arabia took the decision to adjust retroactively to 1st November 1979 the price of its crude Arabian Light by US$6 per barrel, thus bringing its level to US$24 per barrel. With that adjustment, the average price of Arabian Light for 1979 became around US$17.30 per barrel.

and adding a $5 differential for the quality and geographical location. It is interesting to recall that, according to the marketing contracts, the formula of pricing Algerian crude was to take the OPEC price for Arabian Light as a base for the contract price, to which was added those differentials which were to be negotiated quarterly with customers in the light of the markets of both refined products and tanker rates. With the price of marker crude sticking at $14.55 the Algerian oil was judged to become so under-priced that the contracts had to be revised by taking a notional 'de facto' price for Arabian Light and not the official price. With the subsequent adjustment of this latter price into $18 per barrel, the same relationship of prices was more or less maintained.

This decision represented a real departure from the usual pattern of pricing that crude as the OPEC Marker Crude, which so far had been based on a unanimous decision within OPEC itself.[25] However, in the face of the continued tightness in the market, this substantial adjustment in the price of Arabian Light did not help restoring homogeneity in the price structure. No sooner had Saudi Arabia decided to adjust the price of its crude, than other producing countries started adding more or less the same amount on the price of their crudes. Consequently, the price differential between Arabian Light and Iranian Light became about US$4.50 per barrel and the African low sulphur and short-haul crudes became about US$9 per barrel more expensive than the Marker Crude.

25. By the time that price development took place, the book was already in the hands of the printer, and it was therefore too late to analyse its far-reaching effects on the evolution of the OPEC price structure, and above all on the process of OPEC price administration.

PART IV

Some basic issues of oil price administration by OPEC

As soon as OPEC had taken over the pricing of its oil an entirely new set of oil relationships was created. Prior to the oil revolution the main points of strain concerned government/company relationships and the concerted action of member countries in reversing the old status quo, which benefitted only the companies. After the price take-over, the strain is caused by new factors, mostly concerning relations within OPEC and between OPEC and the rest of the world, i.e. the oil importing countries, whether developed or developing.

Many endeavours were made within OPEC to formulate long-term strategies for the regulation of the ensuing new pricing relationships. Best known and important among these was the historic event of the Summit Conference of the Sovereigns and Heads of State of the OPEC member countries held in Algiers in early March 1975. The resulting Solemn Declaration laid down some basic long-term objectives for OPEC to follow in defining its pricing policies. The following are the salient points:

(a) The preservation of the real purchasing power of oil revenues (keeping the price of oil constant in real terms). It was clearly stated in the Declaration that the price of oil must be protected against the eroding effects of world inflation in terms of the transfer of goods and technology, etc. The maintenance of the real price of oil could, in accordance with the Solemn Declaration, be considered as a minimum price adjustment, especially since the Declaration sets out in the other long-term pricing objectives a more ambitious and longer-term strategic planning approach.

(b) The conservation of petroleum. The price of oil should from now on reflect the scarcity value of this depletable commodity, in contrast to the previous situation under the companies in which the movement of reserves to production ratios in the OPEC areas, especially in the Middle East, was sharply downwards in the sense that the net additions to reserves were at a lower rate than the net depletions, i.e. less oil reserves were added than oil extracted.

(c) The valuation of oil in terms of its non-energy uses. The price of oil should reflect the unsubtitutable nature of oil in sectors other than fuel burning, where it could be replaced by other available sources of energy such as coal and nuclear energy. In sectors such as transportation, feedstock for the petrochemical industries and other 'noble' uses, oil cannot be replaced except by unconventional sources of energy such as synthetic oil. Oil prices should therefore reflect the 'noble' nature of this scarce commodity.

(d) Oil prices should be determined in the light of 'conditions of availability, utilization and cost of alternative sources of energy', in the sense that long-term prices of oil should reach parity with the marginal costs of producing a perfectly substitutable synthetic barrel of oil from coal.[1] In other words, oil prices should be planned to increase in real terms up to a level equal to that of marginal cost.

However, the conditions under which OPEC was working its post-1973 pricing policies hindered the systematic implementation of most of those objectives, on which there had been unanimity. Pricing strategies of the Solemn Declaration were intended to be part of a global 'package' for energy and development to be negotiated within an international dialogue. After two years of protracted negotiations, the CIEC, or the so proclaimed North-South dialogue ended in failure. Moreover, the state of the world economy, especially the general recession of most of the OECD economies during 1974 and 1975, had great impact on OPEC pricing decisions during that period. OPEC

1. Nuclear energy and coal are not perfect substitutes for oil because, even with abundant quantities of those sources of energy, it is only the heavier end of the barrel, i.e. fuel oil, that is replaceable. Other ends, especially the lighter ones, such as gasoline for transportation and naptha for petrochemical feedstocks, are not obtainable except from a barrel of synthetic oil produced from coal at exorbitant costs.

accordingly adopted in reality a rather conservative price attitude that did not completely reflect the overall strategic objectives of the Algerian Declaration, and even arrested temporarily the initial trend of the oil price revolution.

Problems of the real price of OPEC oil

One of the most debated aspects of the OPEC post-1973 pricing policies is the preservation of the purchasing power of oil revenues, or the problem of the real price of OPEC oil. What OPEC was actually doing was to hold down the price in real terms compared to rates of world inflation and the eroding value of the US dollar.

One of the problems arising from the discussion of the movement of real prices is the measurement of the impact of world inflation on the price of oil. The erosion in the purchasing power of the oil barrel as measured by the export price index of the OECD countries is much milder than if calculated by measuring imported inflation from the developed countries into OPEC countries by member countries' foreign trade statistics. For some it is only imported inflation, as shown by those statistics, that could be reliably taken as an indication of the magnitude of the real cost incurred by the member countries in importing their development requirements from developed countries. It is rightly held by many OPEC experts that OECD figures for the export price movements cannot be taken as a meaningful indicator of those real costs for several reasons, especially:

(a) The bulk of trade of the OECD countries concerns intra-regional trade (more than two-thirds of total trade of the developed countries is among themselves) which would blur the export price movement as many of their exports are inter-company transactions.
(b) Actual prices paid by OPEC importers are higher than the average import price because of such factors such as discriminatory price practices against OPEC, inflated transportation rates, unusually high rates of expansion of internal demand for imports due to rising revenues, etc.

Those who believe in export price indices published by the OECD or the IMF as a reliable indicator of world inflation contest the reliability of the indices on imported inflation derived from member countries' statistics. These statistics, they argue, reveal extra inflated costs of trade, the result of the inadequate infrastructure and bottlenecks usually reflected in harbour congestions, road insufficiencies, and manpower shortages reflected in unusually high wages, etc. In other words, it is the problems of development inside the OPEC countries that are partly responsible for causing the extra inflation and should not be counted when measuring the impact of world inflation on the price of OPEC oil. There are of course also some technical problems in monitoring 'imported' inflation, such as the non-availability of up-to-date information, the non-uniform nature of trade statistics, etc.

Currency variation and the depreciation of the value of the US dollar is less controversial. Measurements differ according to the basket of currencies chosen, the inclusion or exclusion of the dollar itself as a major currency, the pattern of foreign trade of OPEC countries with the US in comparison with the other developed countries, etc.

The basket of the major currencies of the Geneva I Agreement with companies, is composed of nine currencies without including the dollar itself. There are those who contest the validity of excluding the dollar, since a major portion of OPEC trade is with the US. Other international baskets like the Special Drawing Rights (SDR) include the dollar, but with such a high weighting (33 per cent) that it would conceal an important part of the loss due to currency variations if the actual OPEC pattern of trade were taken as a basis for measurement.

However, even if we take results indicating the least change from those various approaches to both inflation and currency variations, we find a growing gap between the OPEC price and the adjusted price that would be necessary to keep intact the real value of the OPEC barrel. These conservative results could be measured by an aggregate index showing the eroding effects of both variations, and which is constructed on the strength of the OECD export price index but weighted in accordance with the trade patterns of its member countries. It also takes into consideration the movement of the dollar against the basket of nine currencies of the Geneva I Agreement, plus the US dollar, but again weighted in the light of OECD countries' trade patterns. The combination of these two movements yield, according to the index, a cumulative erosion due to both inflation and currency variation in the purchasing power of the OPEC barrel of over 100 per cent since the price adjustment of 1973/74, i.e. the purchasing power of a dollar today is less than half that of December 1973. Accordingly, if the price of $10.85, applied in January 1974, were to be preserved in real terms, OPEC should have undertaken such price adjustments as to bring the price in current dollars to about $23 per barrel. Below is a table comparing the OPEC price for marker crude with the required price in current dollars to maintain the same purchasing power of the OPEC barrel in 1974 dollars (average 1973 = 100).

Year	Inflation Index[1]	Currency Index[2]	Aggregate Index	Indexed Price ($/bbl)	The OPEC Price[3] ($/bbl)
1974	127.00	99.60	126.49	13.64	10.78
1975	139.07	103.02	143.27	15.44	10.72
1976	147.76	97.53	144.11	15.54	11.51
1977	157.00	100.79	158.24	17.06	12.39
1978	162.10	112.89	182.99	19.73	12.70
1979	177.50	117.83	209.15	22.55	17.28

1 Defined as the export price index of OECD countries in terms of national currencies

2 Defined as the change of the exchange rate of US dollars vis-à-vis the nine currencies of Geneva I basket

3 Yearly average of Marker Crude (Arabian Light).

We see from the above comparison that, in spite of the various price increases of OPEC oil in 1979, the average price is still less than the real price in 1974 dollars, by more than 20 per cent. In fact the losses to OPEC resulting from the gap would be almost double if world inflation were measured by the OPEC Import Price Index based on member countries' foreign trade statistics.

The oil price and the world economy

To a great extent the conservative post-1974 OPEC pricing policies were motivated by concern over the world economy after the initial price shock. It was thought that continued upward adjustments of the price, even to take account of world inflation and dollar depreciation, would jeopardize the possibilities for the developed economies to pick up after an arguably prolonged period of economic recession. Holding down oil prices in real terms would, it was thought, alleviate pressures on the balance of payments of the oil importing countries, strengthen the anti-inflationary efforts taken by governments of developed countries, and help achieve sustained growth in the world economy.

Undoubtedly sudden oil price increases will have adverse effects on the balance of payments of oil importing countries. However, as far as most of the developed countries are concerned, the resulting external payments deficits are self-correctable after lead times, which vary depending on the pattern of trade with OPEC countries. Experience has shown that increases in the oil revenues following price adjustments enhance the propensities of OPEC member countries to import more goods and services from the developed countries. It was found that in many cases the OPEC import propensity exceeds unity, in that more than the additional dollars accruing from a price rise would be spent on imports. Even with such OPEC countries which still have limited absorptive capacities to spend on development, the rising trend of imports following the price adjustment indicates higher rates of growth in trade than in oil revenues, and consequently decreasing financial surpluses. The resulting increased trade of OPEC countries, especially those imports for development projects, tends to create ultimately self-correcting factors in the balance of payments of the oil importing countries. The degree of correction depends of course on the distribution of trade among the countries supplying those goods and services, so that countries with a greater export potential to the OPEC areas will have their balance of payments more quickly corrected than other oil importing countries with lesser exports to those areas. Needless to say, the industrialized countries account for the bulk of oil trade with OPEC.[2]

2. The overall OPEC pattern of imports shows that the US accounts for 17 per

The same recycling pattern of the oil money applies to financial flows, since financial surpluses which are not channelled back to the industrialized countries through increased trade are actually recycled through the major international money markets, located, in the main, in those same countries. Whenever this type of recycling occurs, the recipient countries benefit in correcting their external payment imbalances and are able to pay the cost of oil imports, even if their current account (the balance of trade), is in deficit. Here again, the pattern of financial flows plays an important role in correcting those imbalances in external payments. Some countries, like the US, where the bulk of the so-called petro-dollars are absorbed by a highly developed money market, are less affected than other importing developed countries, whose concentration of money markets is weaker. Increasing oil imports into the US would therefore not create real adverse effects on the balance of payments of that country's external payments, because of the high concentration of OPEC surpluses in US banks and treasury bonds, etc.

Oil importing developing countries do not generally enjoy the same benefits as developed countries from increased trade with OPEC. Nor do they have a higher potential for absorbing financial investments. However, for certain developing countries with a high export potential like Taiwan, South Korea, Brazil, the increase in trade with OPEC does partially offset the impact of the oil prices on their balance of payments. But, those countries are the heaviest oil importers among the countries of the Third World because of their high rate of development and growth. In fact, over 75 per cent of total oil imports into the countries of the Third World is taken by only eight of these countries, all with a high level of industrial development.[3]

cent of total imports, compared with Japan which accounts for 15 per cent, West Germany 13.5 per cent, France 8.6 per cent, the UK 8 per cent, Italy 7.6 per cent, the rest of OECD 14 per cent, Socialist countries 3.7 per cent. The remainder is taken up by other developing countries (these figures relate to 1977). It is obvious that within the overall pattern, the contributions of individual OPEC member countries vary enormously.

3. These are Brazil, South Korea, India, Singapore, Taiwan, Turkey, Philippines, Thailand. These countries' average annual net imports in 1978 is estimated to cover 3 million b/d or about 75 per cent of total net oil imports of

Apart from these high export potential countries OPEC's trade with the developing countries is rather limited. For them oil accounts for a small share in their total imports simply because of their low level of economic development and hence modest oil consumption and imports.[4]

Oil price increases also affect the rate of inflation in the world, but to a much lesser extent than is generally thought. Most studies released in the industrialized countries show that the contribution of oil price increases to inflation rates is insubstantial. It is estimated, for example, that a 10 per cent increase in oil prices could contribute an additional 0.3 per cent on average to the inflation rates in the OECD countries. The effect varies of course with the degree of dependence on imported oil; low dependence on oil imports, due to existing domestic energy resources, would imply a low net impact of the oil price on internal inflation. It is pointless to dwell too much here on the causes of world inflation and the part played by oil price increases in creating new conditions in the economic and social structures of the industrialized countries, where internal factors of cost push inflation, like increased wages and continual labour strikes, or demand pull caused by an increased degree of social welfare, are far more important in stepping up the pace of galloping inflation than oil prices.

all developing countries. Apart from India, which is among the countries with the lowest per capita income in the world but which has increasing export potential, most of these countries enjoy a very high level of per capita national income, exceeding $1,500 in 1977 in the case of Brazil and $2,800 in the case of Singapore.

4. Most of these countries have been benefitting from increased OPEC aid that has been extended either bilaterally through government to government loan agreements, or multi-laterally through various regional and international organizations, such as the OPEC Special Fund, the Islamic Fund, the Arab Fund for Economic and Social Development, and also through the national funds in OPEC member countries, like the Kuwait Fund, the Iraqi Fund for External Development, the Fund of the UAE, etc. Most of the aid takes the form of donations (about 80 per cent of total concessionary aid in 1976) or very soft long-term loans. In terms of ratio to GNP, the disbursed net transfer of aid from OPEC countries to the other countries of the Third World accounts for over 4 per cent of the GNP of the OPEC area. If certain deficit countries are excluded, like Algeria and Nigeria, the ratio of aid to GNP of member countries amounts to over 7 per cent. See *The OPEC Aid Record* by Ibrahim Shihata and Robert Mabro.

Finally, the oil price movements undoubtedly have far reaching effects on the terms of trade as effective channels for wealth transfer in the patterns of international trade. It could also have an effect on the growth of the oil importing countries. Exorbitantly high prices in real terms could well affect investment propensities and hence capital formation in the importing countries. It should be stated, however, that such an impact on the GNP growth of the oil importing countries depends on their economic structure and the ability to shift resources to the more productive sectors of their economies. A dynamic economy like that of Japan, for example, which despite its total dependence on imported energy and a very high ratio of oil cost in the total national cost, has nevertheless been able to achieve the highest growth rate in the OECD countries. Japan has been able to achieve fast reallocation of resources to more efficient investments like those in the high technology-intensive industries that have greater chances of increased productivity internally, and hence a higher level of GNP, and an enhanced competitiveness in world markets (hence the persistent surplus in Japanese external payments). Increased investments in such areas of an economy could generate additional income to offset the adverse effects of oil price increases on national incomes.

The problem of OPEC price differentials

Among the important problems which surfaced in connection with oil pricing after the oil producing countries had taken over were crude oil relative values or the oil price differentials i.e. the problem of determining the price relationships between the various types of crudes produced in and exported from OPEC member countries measured against the marker crude (Arabian Light). The purpose was to produce a coherent price structure for OPEC crudes, whereby the transport (short-haul versus long-haul crudes) and quality differences (specific gravities, the API,[5]

5. API stands for American Petroleum Institute, a petroleum scientific research centre, which was the first to set up rules for measuring the differences between crudes, by assigning degrees of gravity (lightness and heaviness) to them. Under this system, higher API degrees indicate the lightness of the

as an approximation of the refinery yield, the sulphur content, etc.) are allowed for in such a way that all OPEC crudes have a relative price exactly equal to their relative value to the refiner at the location of use. With such a price structure, it was hoped that there will no longer be any incentive for consumer substitution between individual OPEC crudes and hence no one crude could be adversely affected more than another by short-term shifts in the demand for total OPEC crudes.

The previous oil pricing system adopted by the major companies was based, particularly as far as the setting of posted prices for tax purposes was concerned, on the existence of an almost straight linear relationship between the yield of lighter ends and the API of each crude. Hence, the companies laid down a constant value differential of 2 US cents for each full degree of API, up or down, in particular for those between 30 and 40 API, without attributing any special value to other quality differences—notably the sulphur content, or pour point, wax content, metal content, salt, etc. For price differentials arising from geographical location the companies used a simple method based on average long-haul freight rates and calculated on the world scale and kept more or less constant irrespective of the changes in those rates, especially at a time when the oil transportation industry was enjoying relative stability.[6]

The major price developments which took place after the Tehran and related agreements led to a total shake-up of the old system of price differentials, but without OPEC at the same time achieving real success in finding a fully-fledged workable system for price differentials. The reason for this convulsion in the system were the enormous changes that had occurred in the structure of the interna-

crude, and lower its heaviness. Thus a crude of API 40 plus is considered to be very light, and a crude under API 30 is very heavy. Those crudes falling in between are considered medium and light crudes. The system of API served for a long period as a gross indicator of the composition (refinery yield) of a barrel of crude. Hence a light crude yields higher proportions of lighter ends like gasoline, and middle distillates (gasoil, Kerosene) than a heavy crude with lower API degree, which yields a higher proportion of fuel oil.

6. Hence a differential of about 40 US cents per barrel was constantly held between the price of Arabian Light ex-Ras Tanura and Sidon, in the East Mediterranean, irrespective of the movements of freight rates.

tional oil market itself, and the impact of newly-emerging market factors on patterns of demand for refined products.

The very fast pace of industrialization and fuel burning in Western Europe and Japan during the 1950s and 1960s had resulted in serious environmental problems which made governments impose increasingly severe restrictions to preserve their environments. Refiners were not allowed to produce fuel oil for burning which had a sulphur content exceeding a certain level. This in turn necessitated either investments in high-cost desulphurization plants to remove the extra sulphur content from the oil, or the purchase of a certain amount of low-sulphur-content oil (the quantity of which is limited when compared with the total production in the OPEC countries) for blending with other oils available to them in order to meet the required products specification. There was more force in these restrictions by the beginning of the 1970s and consequently a certain market premium was added to low sulphur crudes, compared with high sulphur crudes, a premium which reflected the additional cost borne by the refiners in meeting the environmental restrictions. As already explained, the sulphur premium was even further consolidated following the heavy entry of the US into the market as a buyer of increasing quantities of lighter crudes with low sulphur content to meet the environmental restrictions which were already in operation in the home market. The Tehran and related agreements (Tripoli, Lagos), introduced the first change in the system of differentials by attributing a value for sulphur content which was inflexibly incorporated into a system that was scheduled to last for five years.

In respect of specific gravities, there was a great shake-up in the oil system owing to important changes in the pattern of oil consumption. In the post-war period of great technological transformations in economic structures and energy patterns and the fast rate of shift from coal to fuel oil as an energy base to meet these new technological requirements, the growth in oil consumption in Western Europe and Japan was mostly in the form of fuel oil, rather than the light ends or middle distillates. Europe adopted a different pattern from the US in that it did not encourage greater growth in the consumption of gasoline. The very high taxes imposed on consumers of gasoline resulted in a slower growth in consumption of the light products compared with the heavy fuel oil which most Europeans were encouraged to use. The result was a pattern of demand with a greater reliance on the lower end of the barrel. Crudes of the Middle East were more or less in line with that pattern of demand, especially the heavier crudes, the production of which was increasing at a higher rate than the lighter ones.[7] In the US, on the other hand, there was more encouragement for consuming lighter ends than fuel oil because of the importance of the car industry in the American economy, the expansion of which depended on cheap gasoline and the existence of abundant quantities of cheap gas to replace dirty fuel oil as a burning fuel. This resulted in the US product demand being gasoline oriented, and hence a greater demand for lighter crudes.

The pattern of oil demand in Europe and Japan was already changing with the very high growth rates of their affluent economies. This led to greater consumption of lighter ends for transportation and services than fuel oil. Europe and Japan were in the process of gradually catching up with the lifestyle of the US, with less emphasis on fuel oil, until the slow rate of change in the consumption patterns was suddenly accelerated by the 'oil revolution' of 1973/74, in a manner that destroyed the old system of price differentials as administered by the oil companies. The price reactions in the consuming countries and the energy policies aggressively pursued by the governments of importing countries to limit the growth of oil consumption have hit fuel oil disproportionately hard in comparison with other products. The partial conversion to coal as a burning fuel and the various conservation measures taken by those countries have resulted in a substantial decline in demand for the heavier ends. The trend was further enhanced by the economic recession in Europe and Japan,

7. It is very interesting to note that up to 1970 the reserve-to-output ratio of heavy crudes was lower than that for light crudes, i.e. the lifespan of lighter crudes was longer than that of heavier ones. Currently the pattern has been completely reversed whereby the light crudes' lifespan is shorter than the heavier ones.

which by their nature meant that less fuel oil was burnt. This happened at a time when increasing oil purchases were being made for the US, where the demand pattern was already based on lighter ends. World demand and consequently values for lighter low-sulphur crudes stepped up while values for the heavier crudes dwindled.

The post-oil-revolution period was also characterized by instability in market expectations, especially in spot sales, an instability which was enhanced by successive changes in the energy policies of the consuming countries. This dealt another important blow to the old system of price differentials, as it became extremely difficult to establish a stable value relationship between the crudes in the OPEC area, especially when the API linear relationship is used. The API system, although considered as an approximate indicator for the yield of a crude, could not by itself cope with the new quality variables that began to have a strong impact on the commercial value of crudes offered in the market. Furthermore, some changes in the chemistry of OPEC crude oil supply pattern, especially from the new producing regions showed increasing disparities between the linear system and the actual refinery yield of similar ends. For example, the Nigerian medium crude oil of 33.8 API, yields much higher proportions of middle distillates than Arabian Light of 34 API degrees, which means that a higher commercial value is given to the former over and above the higher sulphur content in spite of their API parity.

Thus the differential system as related to the Tehran and related agreements was completely shaken up by the price adjustments in the fall of 1973. What happened when prices were increased by 70 per cent and 140 per cent respectively in October and December of that year was that those percentage increases in the price were also applied to the API differentials and the sulphur content as well as to the geographical differential. Consequently the API differential of 1.5 US cents per full API degree in accordance with the Tehran Agreement suddenly jumped to 6 US cents per full degree API. Similarly the price premium for the low sulphur content crude jumped to 70 US cents per full degree of sulphur content. These jumps in the valuation of the quality differentials were not a reflection of the market as much as the result of an across-the-board increase for the price of all crudes, irrespective of the differences in quality. In the same manner the sudden increase in the price differential due to geographical location was inconsistent with the prevailing conditions in the freight market.

Faced with these non-homogeneous changes in price differentials, OPEC did little to find a workable solution in dealing with the problem that would, at the same time, realistically reflect market fluctuations. What OPEC did after 1973 in its pricing of oil was to agree on a price level for the marker crude—Arabian Light 34 API 1.7 per cent sulphur content FOB Ras Tanura—without unanimously fixing a price for the other crudes in the hope that an agreement would eventually be reached on a formula for pricing other crudes in relation to marker crude. That formula still has to be found. All that OPEC created immediately after the adjustment in prices was a temporary solution against a non-homogeneous one, by attributing 6 US cents for each API degree for crudes above 34 API and 3 US cents per API degree for crudes less than 34 API, i.e. retaining the linear relationship with one amendment.

Market conditions in the wake of the price adjustments disguised the problems of differentials because of the strong demand for crude, which made buyers accept any differential individually set by the producing countries. The problem of these differentials emerged sharply, however, after the major slackening in demand which occurred over the subsequent period, particularly in 1975 when OPEC production fell markedly to less than that of 1974. The contradictions in that pricing system emerged especially in respect of the large differentials for API degrees and sulphur content. As a result, member countries began to correct downward the differentials in response to the new patterns of demand. This was a form which led to a reduction in the value of the sulphur content from 70 US cents to 30 US cents per full 1 per cent; and the API differential of 6 US cents per degree for crude over 34 API fell to 3 US cents per API, i.e. a return to an even linear relationship for all crudes. This system, was not, despite its adjustment, able to cope with the new changes in the oil market. This led to a complete collapse of the relative values of crudes, without a new system of replacing it, so that each member country was free to price its crude around the price of Arabian Light, taking market fluctuations into account.

A problem then arose over the pricing of Arabian Light itself, and its use by OPEC as a marker crude, the pricing of which required OPEC's unanimous decision. Whereas other crudes were priced individually by their producers, the Arabian Light producers found their crude priced by OPEC collectively. This led to repeated demands from Saudi Arabia for the establishment of a system which would oblige all the states to keep a pricing system for all crudes unanimously decided upon by OPEC, or one which would give freedom to all in the pricing of all crudes, including Arabian Light. Effectively, therefore, Saudi Arabia felt it necessary to establish a new system of OPEC pricing. This is why Saudi Arabia in the Doha price decision considered the price of Arabian Light a function of her national sovereignty and not a function that required a unanimous OPEC decision. In a way this was reflected in the two-tier price system and the choice of the majority (the upper tier price-makers) of OPEC members to use a notional marker crude similar to Arabian Light in quality and geographical location to base their pricing decision on. In practice, however, OPEC kept to Arabian Light as the marker crude that required pricing unanimity, in view of what all sensed to be a failure of the mechanics of the then two-tier system, which ended in the summer of 1977 and gave way to a reunification of the OPEC price structure.

Some attempts were made to set up a system to deal with the problem of price differentials: the best known of these was called, at the time, the Algerian Method (Replacement Value Method), a market assessment of the relationship of all OPEC crudes to Arabian Light. The basis of the assessment was to measure the free market fluctuations in the prices of the refined products obtained from each of these crude oils, as compared with Arabian Light. The typical refinery configuration of the major consuming markets, namely the United States, Europe, Japan and the Caribbean, was used to determine the yields of each of the crudes. The value of the various cuts was determined utilizing the share of that cut in the crude yield and the spot product prices in that market. The gross product worth of each crude was then arrived at through adding up the value of the various products, corrected by applying various premia and penalties to cover products with characteristics outside the market specification. Hence, the total value of a barrel of each crude, including Arabian Light, could be established in each market. The quality differential for each crude in each market was arrived at by comparing the gross product worth of that crude with the gross product worth of Arabian Light. By using the export pattern of each crude to provide weighting factors and assessing the geographical location, based on two thirds AFRA[8] and one third spot, and by employing the marker crude price, a single FOB value for each crude could then be arrived at. This method ignores the old system of attributing specific values to API or other specifics, such as sulphur content, freezing-point, wax or lead content, etc. Rotterdam, United States East Coast, Japan, and the Caribbean were chosen as markets in which the value of the refined products was calculated, because these were the only markets in which the prices of refined products, sold and purchased freely, were published without government intervention. The mechanics of this method are as follows:

(a) The average representative refinery configuration in each of the major consuming markets, namely, Europe,

8. AFRA stands for Average Freight Rate Assessment, which is issued by the Tanker Brokers' Panel, an association of six leading brokerage firms in London, and represents the weighted average cost of commercially chartered tonnages (whether on spot or long-term charter basis) as employed in the international transport of oil. AFRA for each month is published at the beginning of the month and includes all charters concluded up to the middle of the previous month. The rates are expressed in the form of percentages of the 'Worldscale' Index which by itself is an assessment of cost per ton in various routes for a 'standard' 19,500 deadweight (dwt) vessel, taking into consideration various cost elements for that particular vessel in each route. At present AFRA is published for six categories of tankers: General Purpose (16,500–24,999 dwt), Medium Range (25,000–44,999 dwt.), Large Range 1 (45,000–79,999 dwt.), Large Range 2 (80,000–159,999 dwt.), VLCC (160,000–319,999 dwt.) and commencing from mid-1979 a new ULCC category (320,000–549,999 dwt.)

USA, Caribbean and Japan, is chosen. Based on these configurations, the typical yield of each OPEC crude is derived.

(b) The spot petroleum product prices in the free markets of these consuming regions are monitored monthly and the average prices for a chosen historical period (say a year or the previous three months) are calculated.

(c) The value of each cut yielded by the crude in each market is arrived at by multiplying the average price of the product by the percentage yield of that cut. This value is then adjusted to meet the market specifications for that market. The 'gross product worth' of each crude in each market is calculated by adding up the values of the various cuts of that crude.

(d) The quality difference of each crude in each market is the result of the difference between the 'gross product worth' of that crude and the marker crude (Arabian Light) in that market.

(e) The geographical location differential is calculated by deriving the difference in the transportation cost of each crude and Arabian Light to each market using the average tanker size used on each route. The transportation cost is calculated on the basis of a mix (two-thirds AFRA and one-third spot) of the freight rates prevailing in the historical period.

(f) The total differential for each crude in each market is the sum of quality difference and geographical locational difference.

(g) The weighted average differential of each crude is derived through weighting the total differential in (f) by the export destination pattern of that crude to the various considered markets. Then the price of each crude FOB is the price of the marker crude (Arabian Light) FOB plus or minus the weighted average differential.

For various reasons, some of which were objective some political, no concensus was reached on the method.

From the objective point of view, there were indeed some problems associated with this method, the most significant of which were:

(a) It relies on a limited number of markets which were not representative of more than 10–15 per cent of the whole international oil trade. The internal market prices are not

available in regular published form, in addition to the fact that pricing of refined products in these markets are regulated by governments, as will be explained later.
(b) Prices of refined products in free markets like Rotterdam, which are formed freely in the light of supply and demand, are marked by rapid and sometimes violent fluctuations. Owing to its sensitivity to short-term movements in supply and demand, prices react disproportionately to the slightest scarcity or surplus in the market.
(c) The extreme difficulty in adjusting properly the value of each cut in the light of market specifications as refiners process a number of crudes to meet their demand and not each crude separately.
(d) The method relies on historical information regarding price movement and crude destination which could not correctly serve as a base for future pricing.

The oil market and OPEC's price administration

One of the major, much debated issues concerning the OPEC price administration is the relevance of market variations to the process of price setting by OPEC. The importance of this issue has been brought into prominence by the current situation, in which there are increasingly huge gaps between OPEC prices and those prevailing in the so-called spot market. The current price of 18 US dollars per barrel (October 1979) for the OPEC marker crude Arabian Light is less than half its price in the spot market. Similarly, in the fall of 1975, OPEC decided on a 10 per cent increase in the price amidst very unfavourable market conditions caused by the substantial reduction in world demand for OPEC oil, (which resulted in severe downward pressures on the price structure and thus created what was called at the time a market 'glut'). These two examples, among others, suggest that, although it usually takes spot-prices as an indication of the market balances and hence a signal of a change in the price relationship, OPEC does not in fact necessarily price its oil exclusively on the strength of that market. This said, the question is therefore: to what extent is OPEC's price administration influenced by short-term market fluctuations? Or to put the question in another form: to what extent does a market, in the true economic sense of free inter-play of supply and demand, exist as a price determinant for the entire world oil trade?

Due to the fact that demand for crude oil is derived from that for refined products (and by the same token the supply of products is dependent on that of crude), the ultimate market place for crude oil can only be traced in its using ends, i.e. the products markets. Therefore, a market for crude oil cannot exist independently from that for refined oil products. Furthermore, world oil consumption depends in the main on international trade in oil, most importantly in crude oil exported from the OPEC countries, whose low levels of economic and social development do not call for high energy consumption. Moreover, more than 85 per cent of OPEC's total oil exports take the form of crude oil and not refined products. Consumers refine imported crude oil at home, where the bulk of the world oil refining capacities are located.

This structural feature of the international oil industry creates a power symmetry in the partnership of the international oil exchanges which would render ultimate demand for crude oil liable to various measures of control by the consumers/importers in the same manner as supply of crude oil can be controlled by the producers/exporters.

Historically, both supply and demand for oil were influenced by various policy measures taken by governments and large energy establishments in the consuming as much as in the producing countries. The control of demand mainly takes the form of various fiscal and commercial policies decided by governments of the consuming countries. Producers can control the supply of crude by pursuing policies pertaining to production and investment as effective means of influencing the volume of crude entering international trade. No less effective a policy tool for the control of supply is the administration of the crude oil price itself by the producers, which can be made independent of demand and supply balances. The level of the crude oil price could have long-term effects on both demand and supply, as low prices in the world market encourage more consumption, whereas high prices would be reflected in the long run by lower levels of demand. Similarly, price levels could influence the supply of oil entering world trade, though with a longer lead time, as higher prices tend to encourage investments in oil in other producing areas, such as the North Sea and Alaska, whereas low prices would reduce the economic viability of such investments.[9]

On the other hand, for consumers as much as for producers, petroleum trade lies within those strategic areas of economic activity that could affect in various degrees economic growth, balance of payments, domestic inflation, employment, etc. For this reason, the petroleum trade is closely related to world power politics and the controllable nature of petroleum supply and demand has certainly been perceived for a long time. It has reflected among other things the continuous shifts in the controlling power between producers and consumers. This controlability has been greatly accentuated by the oil revolution of 1973/74 which dramatically brought into focus the strategic importance of oil in the balance of international political power. As a result, the degree of control through policy measures taken by consumers and producers alike has increased. For both, oil is too important a strategic commodity to be left entirely to the market!

Policy measures for control of the market by consumers have a far-reaching impact on demand. Oldest and most effective among these policy tools is the taxation system regulating internal consumption. Although varying from one country to another in the developed oil importing countries, the oil fiscal structure has resulted in a general pattern of internal pricing for oil products, always set in isolation from the level and structure of the OPEC price. Consumers inside the developed countries, especially those of Western Europe, pay at the delivery end much higher prices for the products they acquire than the

9. Energy investments are not always motivated by criteria of economic viability. Considerations related to national security of domestic energy supplies or those related to balance of payments may warrant such investments irrespective of their economic viability.

income obtained by producers for each barrel exported by them to those countries (whether in the form of world market prices or in the form of government-take). The huge differences in the two price structures are due largely to the taxes and levies raised by governments of those countries.[10]

The gap between the price paid by consumers and that received by producers has changed enormously over time, as posted prices and tax and royalty ratios have changed. In the mid-1960s the price that consumers in Western Europe were paying for the products representing a composite barrel of crude oil lifted from the OPEC area ranged between 11 and 13 US dollars per barrel, against a 'government-take' for the oil producing countries averaging 80 to 90 US cents for the same barrel. The producers' share of the total final price paid by the end-users in that area was not, therefore, more than 8 per cent. After the price and tax adjustments of the Tehran Agreement this share has marginally increased so that, prior to the price revolution of 1973/4, it reached about 13 per cent. This situation has been changed radically following successive price adjustments and the charnges in government-company relationships, including participation arrangements and ultimately government take-over and nationalization. Today the producers' share in the value of the final product has moved up to about one third.[11]

In spite of these increases in the share of producers in the total value of the barrel, the fiscal system in the consuming countries is still such as to create huge gaps between internal and international prices. The purpose of those taxes is either to encourage or discourage oil consumption or to maximize the consumers' share in the economic rent from

10. Another reason for the difference in the internal and external price structures is the companies' profits and different additions in the product value which accumulate throughout the process of transporting the crude lifted by the companies to the consuming areas and its refining, together with the distribution of the refined products, as well as the various costs involved in the handling, storage, internal transportation of refined products, etc.
11. The following diagram shows the evolution in the respective shares of producing and consuming countries in the total average price paid by the end-users in the OECD Europe for each composite barrel of OPEC crude oil.

the trade of oil (which actually should belong in its entirety to the oil producing countries).[12]

The impact of those taxes on demand is enormous. Whilst it is true that short-term price elasticity of demand for petroleum products is generally low, consumers react substantially in the long run against price and market variations. Important price changes tend to provoke with varying lead times, far-reaching reactions on the volume and pattern of demand. Therefore, varied fiscal structures

Components of the price of a
barrel of refined products in Western Europe

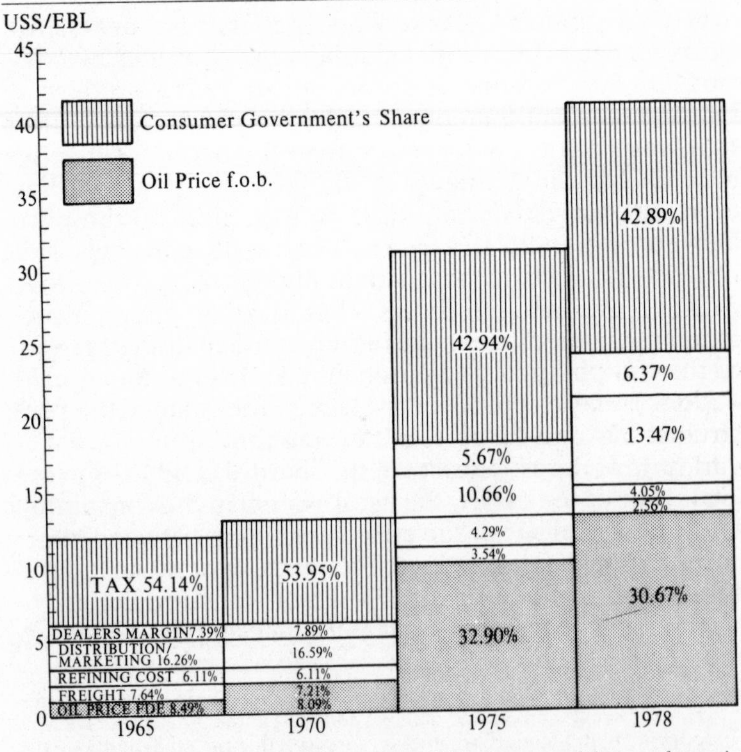

12. This is the case of high taxes on gasoline which, because of the very low price elasticity of demand, are put at optimally high levels that do not affect the economic growth of the consuming countries. Such economic rent that governments of the consuming countries have been reaping in the form of taxes, is usually given the justification that it provides finance for the construction and maintenance of roads (from which the motorists are benefitting)!!

resulting in different internal price levels can have long-term structural effects on demand. For example, very high prices for gasoline in Western Europe caused by high taxes led to relatively low growth rates of European consumption of that product in comparison to the fuel oil which was not subjected to heavy fiscal burdens. In fact, the very high growth rates of European oil consumption during the 1950s and 1960s were due in the main to the replacement of coal by heavy oil products as burning fuel. Very low taxes on the latter products (fuel oil), and hence low internal prices, were essentially behind the exponential growth rates.

With a different economic and energy structure, the United States followed in the past fiscal policies that resulted in different patterns of demand for oil products. Oil served mainly as an industrial input and a fuel for transportation in a country where both chemical and automobile industry represented key sectors in the national economy. Cheap oil was needed for dynamic growth. Being an important oil producer besides being an important consumer, the US did not impose high taxes on the internal consumption of gasoline.[13] This is why this product, together with the other light products, took the major part of the growth of that country's oil consumption.

Until very recently, in fact, the US internal oil price structure was isolated from international prices through various measures of price controls imposed by the Federal Government. Pricing of low-cost internal production in the US was controlled at lower levels than imported oil. The current price decontrols aim at unifying internal price structure.

In the industrialized countries in general quantitative measures to influence demand or trade in oil are not as old as fiscal policies. An exception was, however, the United States. Against its liberalism in internal oil taxation the US

13. It should be stated, however, that in a country like the US, lower prices paid by end-users for all products could generate economic growth elsewhere in the national economy that could offset the economic rent gained through taxation. Nevertheless, the increasing dependence of that country on imported crude oil has been provoking policy movements towards imposing higher internal taxes on the consumers.

resorted to protectionist commercial policies concerning the entry of foreign oil. Unlike Western Europe, where crude oil entered freely without duties,[14] the US imposed tariff barriers on imported crude oil to protect domestic production. More important were the non-tarriff quantitative controls which were rigorously applied from the late 1950s until the early 1970s.

However, the OPEC price revolution of 1973/74 has led to entirely new policies being adopted in all consuming countries. They have moved towards more immediate and direct means of demand control through quantitative measures and mandatory policy controls. Apart from measures aiming at saving energy and putting 'the energy house in order', such as limitations on driving speed, regulations for heating and lighting, regulations for efficiency in energy uses and other conservationist measures, consuming countries are resorting to more aggressive quantitative controls for regulating the volume of consumption, on top of market reactions to price variations. Ceilings on imports in terms of both volume and value are already a common policy practice (the case of France, for example). The IEA's recent policy target of reducing current consumption by 2 million barrels per day is a further step towards firmly controlling demand. For the longer period, the Tokyo Summit Conference of the major industrialized countries (June 1979) has set a new political shift in supply and demand balances, when it set the objectives for oil imports in 1985 to levels equivalent to those of 1978/79. If achieved this will arrest any growth in OPEC production destined for the OECD countries in general.

Conservation policies by consuming countries are currently playing a tremendous role in restricting the growth of demand for oil. In fact conservation was to a certain extent behind the dramatic reversal of consumption trends in those countries since 1973. In 1978 total oil consumption in OECD Europe was less by 0.6 million barrels per day than in 1973, or an average annual decrease of about 1 per

14. For the purpose of protecting its oil refining industry against the competition of cheaper foreign oil products Western Europe imposed custom duties on imported products.

cent.[15] This striking fall in the European oil consumption is to be compared with an average annual growth of over 8 per cent during the five years preceding the OPEC oil revolution and about 12 per cent during the 1960s. Naturally, part of this unusual drop in consumption is attributed to the reaction of end-consumers against the much higher prices paid by them, as a result of both OPEC price increases and the sharp rise in internal taxation. (By 1978, the average total price for an OPEC composite barrel paid by end-users in OECD Europe amounted to about 43 dollars per barrel against a price of 12 dollars paid during the 1960s.) Another factor in the fall in oil consumption is lower rates of economic growth in the industrialized countries which fell during the 1970s to almost half their post-war historical levels. But conservation measures have contributed, according to certain estimates, 20 to 30 per cent of the decline in oil consumption.

Against the consumers' power to influence long-run market trends by controlling demand, producers can no less effectively influence the market by controlling the supply of crude oil entering international trade. The most direct and immediate means of control are the production policies pursued by producers for restricting or expanding crude oil availabilities for exports to world markets. The history of the international petroleum industry shows with clarity that those who are in control of crude oil supplies, whether companies or governments, play a crucial role in

15. Never in any of the last five years did oil consumption in Western Europe reach the level of 1973. The fall in oil imports into Western Europe was more pronounced than oil consumption, mainly because of the increased production from the North Sea. Below is the evolution of oil consumption and imports in Western Europe (in million bbl/d):

Year	Oil Consumption	Oil Imports
1973	15,155	15,405
1974	14,165	14,840
1975	13,505	12,610
1976	14,465	13,725
1977	14,225	13,295
1978	14,600	13,090
(BP Statistical Review)		

determining the shape of the market. A clear case in point was the companies' production 'planning' during the system of complete control and integration of the international oil industry. It was explained in Part I how the total control of the multi-national oil companies crude oil supplies from the OPEC area enabled them to 'programme' crude oil availabilities for export in strict conformity with world demand, which was also, though to a lesser extent, under their control, so that supply/demand balances were continually maintained. We have also seen how, during the period when the companies' control on OPEC supplies was enormously weakened (in the 1960s) the production policies of various producers as well as government/company relationships concerning production levels played an important role in shaping the market (by creating excessive oil supplies).

Finally, the entirely new situation that has been created by the governments' take-over in the producing countries indicates the importance of production policies in shaping the world market. Technical considerations for safeguarding oil-fields have induced many a producing country to reduce production to optimal levels. This was notably the case in Venezuela and Libya which have substantially reduced their production levels to strike a better balance between production and national reserves. In some other producing countries national production ceilings were set by governments with the aim of prolonging the lifespan of oil reserves for better future uses, especially in cases where oil revenues exceed by far the countries' financial requirements. Kuwait set a national production ceiling of 2 million bbl/d after reaching a production level of 3 million bbl/d in the early 1970s. Saudi Arabia, whose production level is the highest in OPEC, has nevertheless chosen to put a national ceiling of 8.5 million bbl/d, which is considerably less than the highest sustainable production levels reached in certain past periods. In fact, with the fast rates of oil depletion in the OPEC countries, there is a growing tendency to conserve oil and to prolong the lifespan of oil reserves by producing only such volumes which, at certain price levels, match the countries' expenditure requirements for development. It is widely held for example in the OPEC area that any level of oil depletion which is not warranted by economic and social development considerations could be considered as a sheer waste of national capital that could be preserved for future development.

Production policies could affect supplies not only quantitatively, i.e., total national production levels, but also qualitatively, by striking better balances among the various types of crude produced by one nation. This is the case of Saudi Arabia, which has chosen to put a limitation on the production of Arabian Light crude in the form of a percentage to the total national production (the 65:35 production ratio) so that more quantities of heavier crude could be moved into the market and more lighter crude could be conserved.

Recent developments clearly indicate the great impact of oil production policies on determining market trends. The new oil production mix of Saudi Arabia had for example created a reversal effect in the supply/demand balance for light products since the first half of 1978, after a long period of so-called market glut. On the other hand, increased production from Kuwait and Saudi Arabia beyond their national ceilings and also from other producing countries like Iraq and Nigeria, effectively supported OPEC's efforts to remove the chaos created in the market following the Iranian oil crisis.

Producers' investment policies for the expansion of production have no less important an effect on the world market, especially at a time when indications for long-term oil shortages or surpluses are increasingly visible. The investment euphoria of the 1960s in the new producing regions, as much as in the old regions, contributed enormously, it is to be recalled, to the creation of the state oversupply during the 1960s. Similarly, the slackened pace of investment in exploration and enhanced recovery in the producing areas during the decade of the 70s is contributing strongly to expectations of oil shortages.

Crude oil producers can also influence demand and the market through price administration. 'Optimized' price levels decided by producers reflect in fact major changes in policies and shifts in controlling power without necessarily being directly influenced by the interplay of supply and demand or variations in cost structure. Our brief analysis of the history of oil price-making has already shown the weakness of immediate linkages between oil price setting and short as well as long-term movements of supply and demand in the market. We have seen that, under the system of complete integration and control of the international petroleum industry by the International Petroleum Cartel, a market for crude did not exist at all. Trade in OPEC oil was no more than inter-company exchanges of crude oil within the almost completely closed circuits of a worldwide integrated system of oil operations controlled by the same companies. No selling and buying of crude outside the companies' internal channels actually took place. We explained how the companies' price administration (setting unilaterally the levels of posted prices) during that period served largely fiscal purposes, mainly determining the taxes and royalties paid to host producing governments. History shows that the companies' pricing strategies were not related to economic forces in the market or relative movements of supply and demand. The successive price cuts which were analysed earlier were motivated rather by the policies of the large energy establishments in the consuming nations (to which the companies belonged) for acquiring cheap energy and raw materials from developing countries.

Even with the subsequent existence of a marginally independent and free market as a result of the weakening of the oil oligopoly, price administration continued to be conducted by the companies but with an emerging partner, OPEC. Again, under the new system, pricing crude oil was hardly sensitive to the interplay of short-term market forces, nor did it reflect long-term economic forces pertaining to supply and demand trends or variations in the cost structure (see above). Also, the evolving market price was more or less pegged to the leaders' price, i.e. the companies' price, which was set around their tax-paid cost plus a profit margin that was considered as 'optimal'. The various phases of the history of oil prices during that period show that the level of the market price as ultimately set by the companies was not a true reflection of the variations in the market or even the balances of supply and demand. It was rather the result of those companies' policies combined with OPEC's actions[16] such as the various policy measures taken by OPEC (including the expensing of royalties, the unification of taxation systems in the producing countries, etc.—see above).

With the OPEC take-over of price administration, market conditions or supply/demand balances started to play a greater role in setting OPEC price levels. Yet the rationale governing political decisions on prices within OPEC more or less reflected short-term political considerations pertaining, in most cases, to the balances in the inter-OPEC interests and requirements (the price 'setters' versus the price 'takers') on the one hand, and the concern over the world economy, on the other. Neither short-term supply/demand balances nor long-term policy objectives were actually determining the price level. A good example, it is again to be recalled, is the movement of OPEC real prices in the post-1974 period (see above).

The analysis above indicates, therefore, that the predominance of policy measures for the regulation of international trade tends to reduce enormously the scope of market forces as a major determinant of the price. The bulk of products' markets in the major consuming centres are, in fact, constrained by policy measures and even structured within a general context of economic policies pertaining to growth in those countries. Crude oil supplies, we have seen, are constrained by the producers' actions and policy measures pertaining to production levels and price administration (which affect both supply of and demand for oil). Consequently, a market in the sense of free interplay of supply and demand for products can hardly be found in most of the important consuming countries. Similarly, a

16. About the administrable nature of OPEC oil prices, see the author's article published in the 'OPEC Review', Vo. II, No. 4, September 1978 *The Administrable Nature of Pricing OPEC Oil and the Impact of New Market Conditions*. See also the paper presented to the First Arab Energy Conference held in Abu Dhabi in March 1979 on *Optimum Production and Pricing Policies* jointly written with Adnan Al-Janabi.

market for crude oil in the same sense of free interplay of supply and demand does not exist in reality, except marginally. The narrow area of free exchanges of both crude and products are meant only to supplement or balance the bulk of the product market, as in the case of the marginal spot market, which reflects the balancing effect on the total market rather than constituting an independent market that could be continually self-sustained (a derived market).

The Rotterdam market is a representative case, whereby short-term balances of the supply of and demand for oil products are more than proportionately reflected in violent price fluctuations. It is in that market, in fact, that marginal barrels of products are offered to marginal buyers, so that the least variations, at the margin, on the supply side as much as on the demand side are translated into violent market reactions. Refiners in the main consuming countries resort to such a free market place to sell products in excess of their internal refinery balances as determined by the patterns of demand for products in the market in which they operate, in the same manner as they enter the same market to buy products in deficit of those balances. This is why the Rotterdam market is highly sensitive to any market imbalance, i.e. when a marginal barrel is offered in the market without a buyer being available to pay the seller's price, this would lead to a more than proportionate fall in the market price level, and similarly, when a marginal barrel is demanded in the market without a seller being ready to supply at the going market price, price flare-ups would consequently take place.

The important factor about those marginal free exchanges of oil products, however, is that they do not represent more than 10–15 per cent of total world oil exchanges. The bulk of the latter is made through established channels in the consuming countries where refiners and distributors perform in 'structured' markets. Prices paid by end-consumers are not the result of free exchanges as much as that of state intervention. Those take the form of not only taxes, various quantitative measures, refining and distribution quotas, etc., but also negotiated prices with traders. In most of the consuming countries, refiners and distributors are not free to set the price of products they sell to consumers as a result of market imbalances (shortages or surpluses). They cannot even automatically pass to the consumers the increases in cost they incur (as a result, for example, of an OPEC price increase or variations in freight rates) before they negotiate with the government.

Parallel to the marginal free markets for products there exists a marginal spot market for crude oil free exchanges. As a result spot sales of crude oil react proportionately to the free market exchanges of products, such as those in Rotterdam so that, whenever a price flare-up in those markets takes place, this would call for additional crude oil purchases for refining and selling as products in the products' free market, and thus a similar price flare-up takes place in the crude spot market. By the same token, when the Rotterdam market is depressed, demand for crude in the spot market is reduced and hence prices for crude oil are lower.

It should be emphasized, however, that violent price reactions in the free market are magnified by the interference of traders, brokers and various middlemen who, for speculative and windfall profit-making purposes, aggravate the state of the market imbalances. Sharp reactions in those marginal markets are furthermore motivated by the short-term stockpiling behaviour of companies and even of energy policy centres in the consuming countries. Moreover, short-term market expectations add enormously to the volatility of the market.

As in the case of free exchanges in products' markets, volatile price movements in the crude spot market are related to a limited volume of total crude exchanges, as it is estimated that of total OPEC crude oil exports, those which are offered in the spot market would not exceed 5 to 10 per cent. The bulk of OPEC oil is traded through channels which reflect more the 'structured' exchanges described earlier. Almost all OPEC crude sales, whether to the ex-concessionaires or to other lifters, are taken by those established users who own and operate refineries and distribution networks in the consuming ends. This includes the multi-national oil companies (whose crude availabilities have been enormously reduced in relation to their downstream requirements), national oil companies in the consuming countries, independent refiners, independent distributors with established supply linkages, etc.

Policy actions on supply and demand by governments and large economic establishments both in the producing and consuming countries have therefore more important an impact on long-term oil balances than the short-term market fluctuations, as reflected in marginal exchanges of crude-oil or refined products. The exchange of crude oil and its cost results in the main from the interaction of those policies that can be determined as a result of the shifts in controlling power among producing and consuming countries, irrespective of the size and cost of oil exchanges. In reality, however, policies are influenced to a great extent by those exchanges, as consumers and producers react to each other as soon as there are important changes in the cost structure both in terms of volume and price.

The spot market could influence decisions on prices in serving as an indicator for the state of short-term supply and demand balances. Very high prices in the spot market indicate a state of oil shortage as much as low prices in those markets indicate a state of surplus. In setting their pricing policies producers usually take into account those market indicators. They could, however, act independently from that market, but only to the extent of setting a floor price which they can sustain irrespective of demand movements. This would mean that in a situation of shortage and market flare-ups, the producers' ability to control the price is enormously reduced as long as they do not dispose of additional quantities to be offered in the market to cope with rising demand. Without OPEC countries being able to produce and offer enough quantities to match the rising demand in the market at the price they set, market strains and price flare-ups will take over the price setting (for higher levels than the floor). Producers would, in this case, be left with little flexibility in controlling the market as was clearly indicated by the oil developments of 1979. It is true that OPEC is currently playing a crucial role in mitigating the effects of the market so that, without OPEC, prices would have risen in general to levels indicated by the spot market. Nevertheless, once OPEC reaches the border-line of its production capacity, it will find its ability to control the market in its entirety somewhat reduced.

OPEC's power to administer oil prices on grounds other than market ones, depends on the degree of supply elasticity at its disposal and the available production capacity in relation to demand variations. This power could be eroded in times of shortages when OPEC's available production capacity falls short of demand (low elasticity of additional supply). In times of surplus or balances between capacity and world demand (high elasticity of additional supply), OPEC has the power to administer the price increases so that it can hold its price structure at the required levels and let its production adjust itself downward in accordance with those price levels.

Future world energy balances and OPEC price administration

A grim picture for the world energy situation has been persistently painted during the last few years. A tremendous flow of forecasts on future energy supply and demand indicate that the world is heading towards a shortage and that the oil crisis is an already endemic phenomenon in the present world energy situation. This state of pessimism has been greatly enhanced by the current situation created in the aftermath of the Iranian oil crisis, when the world has been appalled by the fact that OPEC cannot openendedly cope with the increasing demand. The shut-down of the Iranian oilfields has suddenly brought to the surface the physical limitation of production capacities that OPEC has to face eventually. Despite the substantial increases in other producing countries and notwithstanding the fact that total OPEC production for the first three quarters of 1979 was 5 per cent more than in the corresponding period of the preceding year, the oil market has been under continuous and even increasing strain. This led to a growing concern about likely future energy balances. OPEC oil, it is known, is considered as the world energy 'swing' in the sense that world energy requirements are first met by energy sources other than those from OPEC, and that OPEC oil is to be produced in quantities that are sufficient to fill the gap. Therefore, when the gap widens, world demand for OPEC oil increases, whereas a decline in this demand is expected once the gap is narrowed. Forecasts on the 'required' level of OPEC production varies, therefore, in accordance with forecasts made on world demand for energy and world supply of energy other than OPEC oil. Different estimates were given about the likely world demand for OPEC oil in 1985. Most of them exceed the present production capacity of OPEC.[17]

Against a background of likely imbalances in the future world energy situation, the question is to what extent OPEC price administration can influence world supply/demand balance. OPEC's ability to control world oil markets, we have seen, requires enough capacity for additional oil supplies to be offered in the market at a certain 'administered' price level in the face of a rising demand. Failing such elasticity for additional oil supplies, market disruptions would render any 'planning' approach for pricing OPEC oil extremely difficult.

However, with varying lead times, oil prices can have an impact on the supply/demand balances. Correcting the imbalance by the price mechanism is therefore possible. The central point in any approach that could be discussed in this context is the degree and timing of the reaction of both consumers and energy investors vis-à-vis the price variations. Substantially higher prices in real terms could trigger a long-term self-correcting process in energy markets, as consumers and governments react on the demand side (more conservation, more quantitative controls, fiscal

17. According to forecasts made by Exxon (1978), world demand for OPEC oil in 1985 will be in the order of 40 million bbl/d. The IEA forecast (1978) gave the higher figure of 42.9 million bbl/d. Similar high forecasts were given in 1977 by the US Congressional Research Service (42.8 million bbl/d). The group research conducted in the MIT in 1977 under Prof. Carol Wilson (WAES —Workshop on Alternative Energy Stategies) gave the range of 40 to 45 million bbl/d for 1985. However, some other estimates give lower demand forecasts. The US Department of Energy, for example, made a forecast in 1979 for world demand for OPEC oil ranging between 27.8 and 36 million bbl/d in 1985. The upper limit of this latter range coincides with the forecasts made in 1978 by PIRINC-EPRI.

policies, etc.) and investors in the oil and non-oil energy sectors would be encouraged to react on the supply side. Different approaches could be discussed for the self-correcting process in supply/demand balances as a result of price changes.

One approach is to rely on the market mechanism to re-establish the equilibrium. OPEC could, according to such a market-oriented approach, administer its price on a floor level, i.e. setting a minimum price that it can protect against downward pressures, while leaving the price movement over and above this minimum, or the floor price, to be determined by the market. The history of OPEC price administration indicates that controlling the price by the oil producers is successful only in keeping the minimum price, under which producers were not permitted to sell in times of over-supply. Apart from minor discounts and price cuts that were offered in the market during the periods of 'glut', OPEC was in fact able to maintain its floor price by leaving its production to follow the downward 'swing'.[18] However, in times of shortage, price administration from any 'floor' has no practical significance, whereas any 'ceiling' set by OPEC for its prices cannot resist the upward market pressures. OPEC successfully endeavoured, for example, to put a ceiling on its price (23.5 dollars per barrel) during the third quarter of 1979, but no sooner had further market flare-ups taken place early in the fourth quarter of the same year, than this price ceiling started to wither in the face of new strains.

Consequently, it was often argued that market forces can in fact determine the price, so that market upward pressures resulting from a possible oil shortage in the future would be reflected in much higher OPEC prices than the floor prices, whereas market downward pressures, caused by a surplus or over-supply, would be reflected in lower OPEC prices but only to a level which hits the price cushion set by OPEC as a 'floor'. According to this approach, the

18. It should be noted, however, that the distribution of the downward swing in OPEC was not equal. Total OPEC production in 1975 fell by about 4 million bbl/d from that of 1973. Two thirds of this drop was taken by three member countries, i.e. Venezuela, Kuwait and Saudi Arabia.

self-correcting process of supply and demand would be through a market 'cycle' resulting from the reaction of consumers against the price rise and/or the reaction of the world economy. Thus when oil prices in real terms shoot up violently and successively as a result of market forces, consumers would naturally tend to reduce demand through various measures of government intervention as well as through the natural response of price elasticities of demand. Also it would be possible that economic reaction such as, for example, lower rates of growth and economic activity in the consuming countries vis-à-vis the very high price, would add a further impetus to the decline in demand. The market cycle could also be reversed following the fall in demand to levels that may hit the floor price. In such a case, lower market prices may generate a revival in demand, which is in its turn reflected in higher market prices, and so on. It is thought that such an approach based on purely market mechanisms and natural reaction of demand vis-à-vis price variations would eventually lead to restoring the equilibrium in the market so that the market itself would restore to OPEC its ability to administer its price from a 'floor', while the market would take care of the upper price movements.

This approach does not, however, give enough weight to the structural and political consequences of market price fluctuations. For both producers and consumers, the oil price is too important a factor for economic growth, balance of payments and social change, to be left to the hazards of the market. On the other hand, investments in energy are too costly to be reversible. It is inconceivable that consumers reverse their plans for investment on conservation and/or on other sources of energy, following a market 'cycle'. It is no less conceivable that producers let their development programmes fluctuate in the light of those cycles.

Another approach based on 'planning' by anticipating demand reactions to the price could also be envisaged, as a means of OPEC influencing that process of correction. It could be argued that, instead of leaving the price over and above the floor to be determined by the unforeseeable forces of the market, and the equilibrium between supply and demand to be achieved through hazardous price fluctuations, OPEC can, through the price, influence the market itself. This approach is based on the theory of an 'equilibrium' price that OPEC can set from time to time in accordance with its assessment and forecast of demand variations in relation to its production capacities. When, for example, the rise in demand is expected to exceed the available OPEC production capacity, much higher prices could be set so as to bring demand down to levels that match that capacity. With certain lead times and price elasticities to be carefully observed, higher levels in the price could be a means of pre-empting the market balance. Similarly, OPEC could set a lower price in real terms to reactivate demand and bring it up to the required level matching with OPEC capacity in case of an unusual fall in world demand for OPEC oil. Again through consumers reactions to the price changes an 'equilibrium price', at a lower level could keep or maintain OPEC's share in the world energy markets.

Although in theory this approach presents a more rational framework for achieving market equilibria through price planning, in practice, however, the results obtained from any price 'observatory' cannot be controlled and the same problems arising from the first approach would ultimately be faced.

A third approach for correcting supply/demand imbalances through pricing, which has gained wide acclaim, is that in pricing its oil, OPEC should adopt a pricing path which leads to parity with the cost of alternative sources of energy. Gradual increases in real terms can be 'planned' by OPEC within a chosen time horizon so that, after a certain length of an energy transition period, oil prices would be equalized with the marginal cost of producing energy substitutes. It is often argued that planned increases in the real price of oil would enhance efforts in the consuming countries for more conservation and investments in the alternative sources of energy, so that world balances of energy would be maintained and world pressure on OPEC oil would be alleviated. Such a concept assumes naturally that the future OPEC share in world energy markets would be reduced in favour of other sources of energy.

But, this approach encounters certain intellectual problems related to the definition of the marginal cost of alternative sources of energy and the magnitude and timing of the substitution process. Synthetic oil from coal being the only complete substitute for oil, its cost would presumably be taken as the landmark for future price parity of OPEC oil.[19] One of the problems involved in testing the marginal cost as a basis for pricing oil is that it is hypothetical, in the sense that it is not yet commercially established and that it changes dramatically over time. Even with establishing a commercial value for the synthetic hydrocarbon substitutes there still exists the problem of defining the scale of substitution and its timing. The physical limitations and technological problems resulting from the process of substitution are still to be tested.

Finally, an approach to the problem of long-term supply/demand balances may be sought through creating the right conditions for increased oil availabilities from the OPEC area to cope with world demand. Among other factors, especially those related to the technology acquisition from the developed consuming countries, pricing policies could help enormously in bringing about a better energy balance through expanding OPEC capacities. Oil prices could be set at those levels which would encourage oil producers to invest in intensive exploratory efforts for the search of new oil reservoirs and/or in improving the efficiency of recovering oil from existing reservoirs (thus adding new recoverable reserves from those reservoirs). At the same time, prices may be set at such levels as to encourage consumers not to reduce substantially their dependence on oil in favour of other sources of energy.

Producers are deeply concerned about the fast depletion

19. Coal and nuclear power for electricity generation could replace oil only in certain uses, namely, the industrial fuel, which means that they can substitute the heavier part of the oil barrel. The lighter cuts, such as gasoline, naphtha, etc., can be substituted only by a barrel synthetically manufactured from coal. There are various estimates about the marginal cost of producing that barrel. The current estimates range between 35 and 55 dollars per barrel (in constant dollars).

rates of their oil reserves, especially since the beginning of the 1970s. Recent developments in the reserves to output ratios (a measure for the lifespan of oil reserves) indicate a dramatic fall in the relative movement of reserves replenishment to reserves depletion. For example, the average volume of yearly additions to oil reserves in the Middle East is currently half that of oil depletion (production), whereas during the 1960s the new oil reserves added to the existing ones were about six times the depleted oil.[20] The persistence of those trends involves great economic and political risks, which justify conservationist trends in the producing countries. Consumers, on the other hand, are concerned about the future of oil availabilities from the OPEC area and the conditions for securing oil supplies.

Pricing policies could aim at striking a dynamic balance between the oil depletion in the OPEC area and oil replacement. It is in this area that the prospects for adding new recoverable reserves are greater than in any other producing area of the world. Nevertheless, exploratory efforts in search of new oil are the lowest in the world.[21] Furthermore, any net improvement in the recovery rate of oil from the existing reservoirs can add substantial quantities of new recoverable oil reserves to offset, partly at least, the high depletion rates.[22]

The level of oil prices plays a crucial role in creating enough incentives for OPEC countries to embark on huge

20. Between 1971 and 1978 the total net addition to the Middle Eastern oil reserves was in the order of 30 billion barrels (or an average of 4 billion barrels per year), against a cumulative production of about 60 billion barrels (or a yearly average of 7.5 billion barrels). During the 1960s the total addition was over 180 billion barrels from which about 30 billion barrels only were extracted.
21. OPEC countries as a whole accounted for about 3 per cent of the world total exploratory efforts in 1977, with a share of only 1 per cent for the Middle East. In terms of drilling density, i.e. the number of wells drilled in each 10,000 sq. km., the Middle East has a share of five wells only, against 450 wells in North America and 85 wells in the Soviet Union and 100 wells as a world average.
22. The current estimate of about 450 billion barrels for total OPEC recoverable oil reserves are based on an average recovery factor of 25–30 per cent. An improvement of 10 per cent, for example, which is technically not unfeasible, in the oil recovery (through more secondary methods and enhanced tertiary techniques) would bring an increase of about 30 per cent in the resource base of the OPEC area.

and exorbitantly costly investments for increasing the resource base, especially in generating the additional financial requirements. OPEC current prices (October 1979) would not be enough to solve the financial problems which most of OPEC countries are facing because of the increasing cost of economic and social development of their countries.

However, no matter what approaches are envisaged for the price function in restoring oil balances, one concept for pricing OPEC oil on which universal acceptance is gaining momentum and which could serve as a bare minimum pricing guide is the preservation of the purchasing power of the OPEC price. Period price adjustments could be made to offset the eroding effects of the inflationary trends in the developed countries, as well as the fluctuation of the value of the monetary unit of account denominating the value of oil (so far the US dollar) against the major international currencies. Solutions could be found for the technical problems which this concept could involve in the measurement of the variations, especially those related to the world inflation (whether to take the domestic rate of inflation in industrialized countries or the landed cost of OPEC imports, see above).

EPILOGUE

Energy and development: interdependence among nations

None of the commodities entering international trade is as closely related to world power-politics as petroleum. The foregoing brief analysis of the structural changes in the international petroleum industry, especially those concerning producing/exporting countries, cannot therefore be really understood in isolation from the continuous shifts in world power-politics.

It was by virtue of the oil concessions (and of consortia), all territorially exclusive (full monopoly), which prevailed in the major oil producing countries of the Middle East, that the interterritorial major oil companies were able to control totally the production of crude oil extracted from the region, and hence to dominate, until recently, the world oil trade. But it is common knowledge that this system of oil hegemony by a handful of large transnational firms was the fruit of the direct or indirect colonial domination of the Big Powers over the oil-bearing areas in the Middle East, especially in the post-First World War period. It was then that the corpse of the dead Ottoman Empire was first carved up by the old colonial powers, Great Britain and France, only to be effectively and much more forcefully shared out later on with the emerging great power, the United States of America. When 'agreements' for oil concessions were concluded with the oil producing countries of the Middle East most of the latter were colonies or protectorates under the mandate of the victorious Powers of the old colonialism and had therefore no political free will (sovereignty) in negotiating further. The politically 'independent states' among the producing countries were no less deprived of any right to national sovereignty. Independence was 'given' to them while the real controlling power remained in the hands of the decision-making centres in the metropolis.[1]

For a long time this unusual phenomenon of a high degree of power concentrated in a limited number of highly industrialized countries with American interests acquiring a majority share after the Second World War enabled them to control world money and trade mainly at the expense of the oil producing countries, whose bargaining power was next to nil. Because of the huge imbalance in world power relationships, any endeavours on the part of those countries to emancipate their national wealth from the companies' domination was doomed to failure, as was clearly shown when the Iran of Mossadegh (and before it Mexico, although with different consequences) tried unsuccessfully to reverse this situation in oil relations.

However, under growing pressure from the other partners in the oil trade, this absolute oil hegemony could not be indefinitely sustained. Although developed nations other than the Big Powers were actually benefiting from this lopsided system of oil trade, by acquiring a scarce source of energy almost free, conflicts between national interests and the Cartel pushed them to look for other sources for achieving 'energy independence'. Thus, the first signs of an effective challenge to the oil hegemony were the result of a struggle for oil power among the developed countries themselves. Later on this was reflected in important changes in the industry's structure.

1. Iraq, which in 1931 became an independent sovereign state and a member of the League of Nations, had to pay the very high price of replacing the agreement with the Turkish Petroleum Company (concluded in 1925 when Iraq was still under the British mandate) by the 'agreement' with the Iraq Petroleum Company (the same group with a new name): British colonialist power had threatened to remove the Wilayat of Mosul from the country's territorial sovereignty. As was explained earlier in the text, the former agreement confirmed the principle of relinquishment of unexplored and undeveloped plots of land to be returned to the government for free disposal (and hence for possible competition from other foreign companies developing oil in Iraq). Above all, it was not territorially exclusive (the system of plots). The 'agreement' with the IPC group abolished the principle of relinquishment and confirmed the territorial exclusivity since it, together with two other subsequent agreements, covered the entire territory of Iraq.

ENI and its imaginative Mattei were effectively the first to signal a possible shift in the oil controlling power.

This happened at a time when the tide for political and economic emancipation of the poor raw-material-exporting countries began to gather momentum. The days of the old form of colonialism were dramatically ended by the Suez Crisis and a new era of world power-politics began to dawn. It was therefore no coincidence that the first signs of a 'weakening oligopoly' were visible only in the aftermath of that crisis and it was, likewise, no coincidence that OPEC came into being in Baghdad shortly afterwards (1960). With the emergence of the Third World and Non-aligned Powers the first, albeit very modest, shift of the controlling power in the world system of international trade in raw materials in favour of its producers/exporters was already on the horizon.

In spite of the timidity of its actions during its first decade OPEC was instrumental in ushering in a change in a situation in which its members had been, for decades, the real victims. But the really revolutionary years of the oil industry were those of the 70s, during which oil power shifted dramatically towards the OPEC countries. It was the interaction of the oil policy actions taken individually by the organization's members, as well as those taken collectively, that really caused this reversal.

It is now universally admitted, even in the West, that prior to the oil revolution of 1973/4 crude oil was made available in world markets by the major oil companies at such artificially low levels that the oil producing countries were actually 'subsidising' the post-World War II 'miraculous' economic growth of Western Europe and Japan. But this very expensive 'gratuity' was taken from poor countries for whom oil represents virtually the only resource for development and for economic and social structural transformations. In fact, the total control of the oil companies over the development of oil and its pricing enabled them to continually impoverish the oil producing/exporting countries in favour of the industrialized importing countries, by successively undercutting the price of the OPEC (Middle East) oil. By its creation, OPEC was able to put an end to the process of erosion of its members' revenues, but only in nominal terms. Its success in freezing the oil price throughout the whole of the 1960s meant further erosion in real terms of its revenues, and hence a further deterioration in the terms of trade of the oil producing countries. To make things worse, the old concessions and the fiscal relationships between the operating companies and the oil producing countries were such as to allocate to the producers only half of the declining 'posted price'. It was that shrinking per barrel 'government-take' (about 80 US cents until January 1971), and not the posted price or the market price, that represented the real cost of acquiring a barrel imported by the consuming countries, especially those to which the multi-national oil companies belonged[2] and consequently represented the share of the producing countries in the huge rent from this scarce commodity.

The irony was that the 'subsidy' for economic growth in the developed importing countries was taken from a natural resource which is depletable and which has a limited lifespan no matter how prolific the oil reservoirs. A barrel extracted is lost forever and cannot be replaced by another barrel, except perhaps by high risk investment in search of a new oil discovery which may or may not prove fruitful, or by complicated and expensive techniques of enhanced recovery from existing reservoirs.

There were several adverse results of such lopsidedness in the oil partnership. Some were far-reaching, not only for the life of the oil producers, but also for the world energy balance. One main result was that the rich became richer by acquiring, almost free of charge, a strategic commodity vital for the technological and social transformations of modern societies, whereas the poor became poorer as their plans for economic and social developments continually shrank, or were arrested. As the needs grew in the oil developing/producing nations and the purchasing power of the

2. For other importing countries the cost of acquiring the OPEC barrel was higher than the tax-paid cost incurred by the companies, as it was nearing the 'market-price' (which during the latter part of the 60s was around half-way between the tax-paid cost and the posted price, i.e. around US$ 1.30 per barrel).

barrel shrunk, the oil revenues of most of them were barely enough to meet their day to day foreign exchange requirements, and thus relegated long-term investments for structural diversification to a second level of priority for national expenditure. Secondly, the extraordinarily low levels of the acquisition cost of oil by the developed importing countries led to a huge wastage of energy, as many conservation measures consumers could have taken would have made them incur a higher cost than that of the consumed oil.[3] With abnormally cheap oil, the incentive for consumers was to waste rather than save energy. Thirdly, OPEC's oil reserves were depleted so fast that their lifespan was dramatically shortened, in spite of the fact that oil discoveries in the producing areas, especially in the Middle East, were tremendous.[4] Lastly, these fast rates of oil depletion were detrimental not only to the oil producing countries, since the replacement cost to them of this very low cost oil (through exploratory efforts and enhanced recovery) has risen in a prohibitive manner, but also to the world-wide long-term energy balances. It is these high rates of depletion that are behind current uncertainties about OPEC's capacity to meet the world's future energy requirements.

Maintaining the old status quo of the companies' domination over the oil trade and the unequal distribution of wealth generated from oil would have simply meant that the oil producing countries would be left impoverished, without development and energy resources in future, while the industrialized countries would be able to achieve their growth and energy transformations.

OPEC's real revolutionary achievement was not so much in its price action as in triggering the process of dynamic

3. It is known, for example, that the cost of saving energy through investments in insulation of buildings (or electrical installations) that would lead to lesser per unit consumption of energy, were higher than the cost of consuming energy.
4. Between 1957 and 1977 oil reserves in the OPEC area of the Middle East and Africa multiplied by more than three times. However, those areas' production capacity measured in the reserves to output ratio declined during the same period from a peak of over 130 years at the rate of production in 1957 to about 35 years at the rate of production in 1977.

change towards completely taking over from the major oil companies all aspects of their national oil industries and the policies affecting them. It is the restoration of the lost right of national sovereignty over their vital resources and the effective exercise of that right which really counts in assessing OPEC's balance sheet during those revolutionary years. The pricing takeover by the historic OPEC meeting of October 1973 in Kuwait was only one major part of this dynamic process. Other parts were perhaps more important in achieving the real shift in the balance of world oil power. The termination of oil concessions and the government take-over of oil operations, whether by nationalization or otherwise, represented the real step towards the emancipation of the national oil wealth and thereby asserting the producing countries' oil independence. No less crucial was the gradual replacement of the major oil companies by the OPEC national oil companies as the predominant force in the world crude-oil market.

Now that the system of oil concessions has totally collapsed, and the major oil companies have lost the power to control the world trade in crude oil, a new status quo of oil power relationships allows a fairer world distribution of wealth generated from energy. More than that: the current oil relationships herald a new era of international economic relations, whereby a lesser power asymmetry will gradually replace the old relationship of domination between raw material importing/developed and exporting/developing countries. In other words, the experience of the oil exporting countries in changing the old status quo could be considered as the spearhead for a new international economic order.

However, with the fading role of the oil companies as the intermediaries of energy exchange, producers and consumers have to face each other directly under a new set of rules in a new oil 'game' for the distribution of world economic growth and welfare.

Sine qua non for development though it is, economic independence cannot by itself lead to that development in isolation from the world system of exchanges in which the partners in international trade have to coexist within a complicated framework of national and international conflict of interests. By controlling the oil industry, the companies used to dominate the flow of exchanges necessary for development in the oil producing countries. Now, in contrast, the countries are no longer dependent on the companies and therefore enter a world of interdependence between producers and consumers in which they themselves must operate optimally these exchanges. Consumers need producers for oil supplies to the same degree that producers need consumers for trade and development. Expanding the value of exchanges between these two groups would simply and ultimately mean expanding welfare. Therefore interdependence and mutual needs should in the long run create the required structured and stabilized channels for those exchanges. However, the relatively sudden reversal of the old situation has created new problems, the solution of which requires international cooperation, and hence dialogues. This will need vision and a long term strategy on the part of all partners in the oil 'game'.

Three sets of problems must be highlighted; the relationship between OPEC member countries, the relationship between OPEC countries and the developing countries, and the relationship with the industrialized countries.

Inter-OPEC relationships

In dominating the world's oil trade, the major oil companies regulated oil relationships not only among consuming/importing countries and producing/exporting countries, but also those between the latter countries themselves.

In taking global decisions on virtually all the vital aspects of the oil industry concerning all the producing countries where they were operating, the companies actually for a long time decided on behalf of each producing country the policies pertaining to investments, marketing and pricing of its oil. Those decisions, covered investment programmes for future capacities, actual production levels, market shares, the pricing of crude, the price differential among various crudes exported from the producing countries, market destinations of those crudes, percentage sharing of various crudes to cope with refinery configurations in the consuming areas, etc. This inter-territorial approach for conducting oil operations meant ultimately that those companies actually passed the territorial sovereignty of each of the countries concerned. Without this system of horizontal integration, they would not have been able to control the world's oil trade in such a perfect manner as to serve their own interests at the expense of the producers. It must be stressed that almost all of those companies possessed complete information about the geology of the entire area, its reservoirs, its capacities, its potential for expansion, its technical problems, etc. It was on the strength of such global and regional control of information that the companies' 'territorial' oil decisions were made, not only on investments and production levels, but also on such other, no less vital, aspects as marketing, pricing, blending of crudes, etc. In a sense, therefore, the existence of the oil cartel in the OPEC area was a factor of integration of its oil operations but in a way which served the companies' interests at the expense of the oil producing countries by reducing their sovereignty to the collection of taxes.

With oil independence, a new set of intra-OPEC relationships has been created, whereby a strict 'territorial' approach in dealing with the national oil affairs is suddenly replacing the old inter-territorial pattern of the oil companies. The horizontal integration linkages no longer exist. It is true that the existence of OPEC is a factor of integration among its members, but only in so far as the determination of the absolute level of the price of oil is concerned. All other aspects of the industry, especially those pertaining to investment programmes, production capacities, marketing, market shares, etc., are conducted on a purely national basis without real coordination being sought among the member countries bilaterally or multilaterally with a view to reducing the risks of the oil operations and maximizing the economic benefits of the nations concerned.

It is of course inconceivable that the old approach of inter-company horizontal integration be applied among sovereign states. For the producing countries oil is so central in their economic and political life that it is only natural that with oil independence this vital source becomes part of the state's national and global strategies. From this point of view oil nationalism is a corollary of oil independence; especially in that, for developing countries, like those of OPEC, concepts of the costs and investment risks are related to economic and social development, the benefits of which are measurable in terms of social welfare, and not in terms of purely commercial profit and loss criteria. In this context, it should be recognized that the OPEC area is far from representing a homogeneous group of nations. The oil producing countries differ enormously in their economic, social, political and financial structures. They also differ in their oil potentials. Their national interests should by definition differ in such a manner as to weaken the links of horizontal integration, since national priorities in each country are a function of its domestic economic and political requirements.

However, a tendency to 'excessive localism', whereby each state plans all aspects of oil policy in almost total separation from the other producing states, could increase the economic and political risks of oil independence. It is true that independence enhances consciousness of national sovereignty. But a rational implementation of the concept of national sovereignty would justify a partial relinquishment of that sovereignty through cooperation with the other states in order to maximize national gains. An extremely narrow interpretation of the concept of national sovereignty and oil nationalism would ultimately reduce oil independence into a factor for disintegration of the oil operations of the producing countries. It is, therefore, necessary, if not urgent, to mobilize political efforts to overcome the difficulties impeding the coordination of oil policies that have already been discussed in the early sections of this book. Modest, but realistic efforts are much more conducive to 'integration' than over-ambitious programmes that have little chance of political acceptance. Four main areas of cooperation are to be highlighted.

(1) Price differentials. One of the most important threats to

the OPEC price structure is the lack of agreed rules for determining the price of all crudes produced by OPEC in relation to the Marker Crude. Determining the price of only the latter by OPEC without agreeing on the price differentials for the other crudes may cause, especially in times of shrinking demand for OPEC oil, a possible disruption of inter-OPEC relationships, since it would mean ultimately an unequal distribution of the cuts in OPEC production and market shares in response to the market. It is through a homogeneous and consistent OPEC price structure that the respective shares in the market can be monitored in the light of the *constant movement* of demand in relation to OPEC's capacities to produce various crudes, and its distribution among member countries. In this way, any 'swing' in OPEC production would be spread proportionately among the producers through a price mechanism that would realistically reflect the real value differentials attributed to the differences among the crudes in terms of quality and geographical location. Leaving the price of all crudes, other than the Marker Crude, without unanimous OPEC decisions would, in a time of market 'glut', subject the Marker Crude to increasing pressures and make it absorb a major part of the downward production swing.

OPEC's past success in living with this problem without solving it would not necessarily mean that its price structure would survive possible future pressures equally well, as in the downward swing producers may ultimately react through price changes for the redistribution of market shares. No less serious a source of price tension is that without a solution to the problem of price differentials, those producers would see no point in continuing to have the price of their crudes decided by OPEC without having the same pricing freedom as the other producers.

The problem of price, has been and will continue to be the raison d'être of the very existence of OPEC, a fact that renders the solution of the price differentials a crucial policy priority for the survival of OPEC itself.

It is obvious that in times of shortage, as in the market situation prevailing from mid-1978 through 1979, the problem of price differentials is not as serious as in times of market 'glut'. The upward market pressures would allow differentials to go as high as buyers are willing to pay. In this context there must, however, be a distinction between the willingness of buyers to pay higher prices in reflection of the market shortage on one hand, and their evaluation of the real value differential reflecting the quality and geographical location differences on the other. Failing to agree on the problem of differentials in such situations would, therefore, be no less harmful to the long-term interests of OPEC as in the case of falling demand in relation to available capacities.

(2) Crude oil marketing. Cooperation among the national oil companies of OPEC countries is no less important in this area than in the pricing although both are inter-related. The present situation of almost totally uncoordinated marketing policies could, especially in times of a buyer's market, further strengthen the bargaining position of the buyers and thus add to potentially harmful competition among producers. Such a coordination has become all the more urgent with the fast shrinkage of the major oil companies as the main marketers of crude oil in favour of the national oil companies. These latter are becoming an effectively dominant force in the market that determines not only the respective market shares but also plays a major role in shaping the market itself and even setting the price leadership. The success of the major oil companies, one has to recall, in maintaining their fictitious price system, including a fictitious system for price differentials, was made possible only through the various formulae of market sharing. Obviously OPEC cannot resort to the same formulae as its marketing policies reflect the requirements and strategies of states and do not correspond to the commercial behavior of an oligopoly. Still, some measure of market coordination is possible to strengthen the leader's role of the national companies in world markets. Exchange of information concerning the sales contracts, customers' performance, market quotations, etc., would strengthen each national company's bargaining position without jeopardizing member countries' marketing strategies. Nor would

such cooperation amount to oligopolistic practices of sharing the markets.

(3) The upstream operations and investments. Since their assumption of full management responsibility for the upstream in their respective countries, the OPEC national oil companies have been successfully conducting this rather technically complicated phase of the industry, without meeting any real difficulties. Operationally, therefore, the necessity for OPEC's national oil companies to coordinate their programme has not yet arisen. Such necessity will, however, arise in the future as the operational difficulties increase, especially in maintaining the current levels of recovery efficiency, and as the need to expand capacities to meet world demand grows.

No matter what is said about the possible structural changes in the energy patterns of the consuming countries and the policies of reducing dependence on OPEC, oil from this area will continue for years to come to be the main source of energy entering the international trade. OPEC will therefore be called upon not only to maintain its current production level, but to meet the incremental world oil demand.

It is known that most of OPEC member countries are producing at or near capacity. Maintaining those levels over time would create huge technical problems and involve exorbitant rising costs in highly risky investments for the maintenance of oil reservoirs and enhanced recovery. For the few other countries with growth potential, the financial and technical risks of investment in the expansion of capacity are even greater. Cooperation among member countries in those areas may therefore eventually be caused by the need to mitigate the increasing technical and financial risks resulting from the present total dependence of each producing country on foreign companies for providing the required technical expertise. Even the political risks of such dependence may lead the producers to improve cooperation in implementing their investment programmes and in coordinating their contacts with the foreign companies, as they may find themselves eventually replacing their old dependence on the major oil companies in the oil trade by a new technical dependence on foreign supplies. Through cooperating among themselves, the national oil companies could well achieve a reasonable degree of self-reliance and could at least improve their bargaining position vis-à-vis the foreign suppliers.

Cooperating in the area of exploration, drilling, geological assessment, geophysical surveys, etc. would be very helpful in maximizing the gains of each national company. The exchange of information in this area would be even more instrumental in reducing investment risks. Similarly, in the area of enhanced recovery, including water injection, gas injection, steam injection, etc. and the exchange of information and experiences over these matters would help the national oil companies enormously in overcoming the increasing technical difficulties and rising costs. In the context of regional joint action, OAPEC's experience is one of a pioneer and its success will certainly lead to creating the right atmosphere for more cooperation among the national oil companies. The Arab Petroleum Service Company, which was established in 1977 and which has already engaged in drilling activities in a number of member countries, is a step in the right direction. Like many other joint ventures established by OAPEC, such as the Arab Maritime Company for oil transportation, the Arab Petroleum Investment Company, etc. The Arab Petroleum Service Company was meant to be another vehicle for cooperation among the Arab oil producing countries, both in the upstream and downstream stages of the oil industry operations. The intention of OAPEC to establish a new Arab joint company for petroleum engineering work and technical designs is also a positive step towards fostering cooperation on a regional level.

Moreover, as developing countries with relatively fresh involvement in the oil industry, most of the OPEC countries suffer from acute shortages in technical and human resources. Joint programmes for the development of those resources would secure faster and more efficient results and could therefore contribute to reducing dependence on foreign expertise. Again OAPEC's experience in the area of regional cooperation in training promises to promote further cooperation among producing countries.

(4) Oil conservation. From the point of view of the world energy balances, OPEC is generally considered to be the 'residual' or swing energy supplier, ready to fill any deficit between the energy requirements and energy (including oil) supplies from non-OPEC sources. This is why OPEC's production shows such swings, up and down, in response to the variations of such deficits, both in terms of quantities and in timing. As a result, what exclusively determines OPEC production levels is the world demand for its oil, as reflected by lifters' decisions and policies on oil supplies, to meet not only consumption requirements but also the movements of stock build-up.[5]

With the continual high rate of depletion of their resources, OPEC countries will eventually have to adopt common conservationist policies that will actively influence member countries' production levels, and not rely exclusively on the world markets to determine those levels. Such policies are already implemented in certain member countries like Kuwait, where the national ceiling of 2 million b/d is well below the peak production levels reached in the early 1970s. In spite of its very high production levels, Saudi Arabia is also adopting a ceiling of 8.5 million b/d which can be considered as relatively low in comparison with its oil potentials.[6] Other oil producing countries are also heading towards more clearly defined policies for conservation.

It is imperative, however, that some common guidelines are adopted for defining long-term production policies which will reflect the basic concern of OPEC countries to

5. It is interesting to note how during recent years the variations in OPEC production strikingly reflected the movement of stockpiling in the consuming countries. In periods of heavy stock replenishment, as was generally the case in 1979, OPEC's production increases were much greater then increases in consumption, whereas in times of stock depletion, as was the case in 1977 (and partly 1978), OPEC production was below consumption levels.
6. Following the Iranian oil crisis and the resulting market shortages, Saudi Arabia has been producing beyond its national ceiling (9.5 million b/d). Kuwait has also surpassed, for the same reason, its national ceiling by about 0.5 million b/d).

preserve their depletable asset for future generations. Such guidelines would also help the producers to influence market conditions and to counter-balance the adverse effects of the fluctuating movements of world demand, including consumers' short-term policies of stock build-up and long-term policies of conservation. On the other hand, as the oil reservoirs age, more technical conservation measures are needed to preserve them. Cooperation among OPEC countries could be very helpful in overcoming such technical problems in conservation.

OPEC's relationship with other countries of the Third World

As developing countries, members of OPEC are an integral part of the Third World. For them, as much as for any other developing country, accelerating economic growth and social development are of real concern and are motives for their global strategies. Similarly, changing the international economic order so that it provides the real means to accelerate development is as important for any oil producing country as for the other developing countries. OPEC's solidarity with other developing countries should, therefore, be central in its strategies for development. What OPEC has done in changing the pricing system of a particular raw material is serving as an effective instrument for changing the international economic order as far as other raw materials are concerned. Its pricing action should eventually serve as a model for correcting the prices of other raw materials. Consequently, by acting on the price issue, OPEC has effectively strengthened inter-Third World solidarity.

However, facts about OPEC's relationship with other countries of the Third World have been generally obscured, especially in industrialized countries. False political conclusions about this relationship also stem from some basic misconceptions about the problems of energy and development of the Third World nations. The impression created is that OPEC countries are no longer part of the Third World, as the oil price increases have made them the rich among the poor. Those oil price increases, it is wrongly said, are behind the deteriorating pace of economic development in the developing countries, if they have not already resulted in strangling their developmental efforts. Misconceptions of this kind are only matched by a blackout on OPEC's record of aid for development in comparison to the aid performance of the industrialized countries.

The fact is that for the West OPEC's price action has been an easy political alibi to absolve its historical responsibility for development problems of the poor nations. The real causes of the Third World's plight, which stem basically from the nature of the world economic order, have been suddenly forgotten in the face of a huge 'information machine' shifting to OPEC shoulders the responsibility for the development problems of the Third World nations.

Between 1974 and 1977, OPEC's total disbursements to the other developing countries was about 36 billion dollars, or 40 per cent of the total of public aid in favour of those countries, excluding that from the socialist countries of Eastern Europe. The OECD countries, whose Gross National Product is 16 times as much as that of the OPEC countries, provided the remaining 60 per cent.[7] The bulk of OPEC's aid disbursements (about 76 per cent in 1977) took the form of concessional flows, mostly grants and donations, low interest-bearing loans, long grace periods,

7. The following table shows the evolution of OPEC flows to other developing countries and its share in the total public aid for international development. (Aid figures for the period 1974–6 are taken from UNCTAD, and those for 1977 as well as the figures for OECD aid are taken from OECD). The table excludes the Socialist countries of Eastern Europe which contributed $11.5 billions of public aid to developing countries during the same period.

($ million)

	Total OPEC Aid	Total OECD Aid	Grand Total	OPEC as % of Total
1974	7,561	11,302	18,863	40
1975	11,457	13,587	25,044	46
1976	8,978	13,665	22,643	40
1977	7,588	14,696	22,284	34
	35,584	53,250	88,834	40

etc.[8] In terms of the ratio of aid to GNP, therefore, the comparison of OPEC's aid performance with that of the OECD is a striking example of the failure of the industrialized countries to fulfil their commitments to the Third World nations. Though they are developing countries themselves with increasing financial requirements for development and hence with less capacity for aid, oil producing countries nevertheless, have more than proportionately faced the political responsibilities of aid for international development. It is ironic to note that, against an OPEC ratio of aid to GNP that has consistently exceeded 10 in the main donors among the Arab oil exporting countries, OECD's aid to the developing countries, as a ratio to its GNP, never exceeded an average of 0.33. In the case of the richest among the rich, such as the USA, Japan, Germany, Switzerland, etc., the ratio is even lower[9] than the already extraordinarily low level of the OECD average, and has generally declined during recent years. This is why, in terms of aid to GNP ratio, the Arab oil producing countries, especially those from the Arabian peninsula, have been consistently at the top of the list of the world's top 10 donors during the last years, followed by the Scandinavian countries and the Netherlands.[10]

8. The ratio of OPEC's concessional aid was lower in previous years, but with a rising trend as shown below (flows are in million $)

	Total Flows	Non-Concessional Flows	Concessional Flows	Concessional as % of total
1973	1,591	446	1,145	72
1974	7,561	4,057	3,504	46.3
1975	11,457	5,984	5,473	47.8
1976	8,978	3,739	5,239	58.4
1977	7,588	1,847	5,741	75.7

Figures are taken from UNCTAD for 1973 through 1976, and from OECD for 1977.

9. In 1977 the ratio for the total OECD area was 0.31; 0.22 for the United States, 0.19 for Switzerland, 0.21 for Japan, 0.27 for Germany, 0.37 for the United Kingdom, 0.60 for France, 0.85 for the Netherlands and 0.84 for the Scandinavian countries (source OECD).
10. The table below, taken from OECD, *Development Cooperation Review* 1979, gives the Top Ten Donors in the world, in terms of concessional disbursement to GNP ratio. (The OPEC countries are underlined.)

No less important a fact in respect of aid relationships is that more than half of the OECD aid to the developing countries is tied to the purchase of products and services from the donor countries; a fact which further reduces the real value of aid, as in the absence of competition, prices for such purchases tend to be higher than in the international market. The net disbursement of the industrialized countries' effective aid is, for this reason, less than the nominal value of aid.[11] OPEC countries' aid, is freely disbursed without any link being made to exports to the other developing countries. Their main exports to the developing countries, i.e. oil, are generally valued in accordance with official prices. Moreover, in terms of economic growth, donors from the developed countries generally benefit from their aid to developing countries, not only in export-tied aid,[12] but also because much aid is recycled back to the donors in payment for more goods exported to the recipient countries. The recycling of aid helps maintain the growth of the export sector of the donor country by creating and maintaining external demand for it, especially in cases where such sectors suffer from low capacity utilization. In a sense, therefore, aid by the OECD countries could be considered a contributing factor to their own growth, and not a real wealth transfer. Unlike the developed countries, OPEC

1975	1976	1977	1978
Qatar	United Arab Emirates	United Arab Emirates	United Arab Emirates
United Arab Emirates	Qatar	Kuwait	Kuwait
Kuwait	Saudi Arabia	Qatar	Qatar
Saudi Arabia	Kuwait	Saudi Arabia	Saudi Arabia
Libya	Iraq	Sweden	Sweden
Iraq	Iran	Holland	Norway
Iran	Sweden	Norway	Holland
Sweden	Holland	Libya	Libya
Holland	Norway	Denmark	Denmark
Norway	Libya	France	France
	France	Canada	Canada
			Belgium

11. An UNCTAD study carried out in 1967 estimated potential price differentials in non-project aid-financed chemical imports into India at about 29 per cent. See the UNCTAD report on OPEC aid: A Summary published by the OPEC Special Fund, July 1979.
12. *ibid.*

countries do not benefit from the same recycling of disbursed aid to the other developing countries. For them aid represents a net resource transfer without any counterpart (in terms of growth) being obtained. In this sense, OPEC aid is a real sacrifice of development opportunities in favour of other developing countries.

Furthermore, the wealth transfer from the OPEC countries is mainly the result of their temporary financial surpluses, which in essence are no more than the monetary counterpart of the exhaustion of a non-renewable mineral asset that could be conserved for better future uses. In other words, OPEC's aid is taken from a national capital resource, which is extracted beyond the owners' needs, and is hence lost forever. It is because of their wider obligation vis-à-vis the world economy that some of the oil producing countries, especially those of the Arabian peninsula, are producing in excess of their capacity to absorb additional income for development. The disbursement of the resulting surpluses would mean the loss of an asset that is badly needed for future development. Aid from the industrialized countries on the other hand is basically the result of excess production of renewable goods and services and is, therefore, taken from reproducible wealth. For these reasons, aid disbursements would create, in the case of OPEC, long-term adverse development effects, whereas in the case of the developed countries, it can produce positive effects on the growth of the donor countries.

Developing countries' energy and development problems differ widely and so, consequently, would the impact on their development of oil price increases. OPEC aid, in favour of such countries, takes account of these differences.

Because of their high level of industrialization and development, only a limited number of relatively well off developing countries account for the bulk of the total oil imports into the Third World. In fact, about two-thirds of those imports are taken by eight developing countries, with a per capita national income of about $1,000 and above,[13]

13. In 1978, Brazil, the Republic of Korea, Turkey, Taiwan, Cuba, Singapore, Jamaica and Chile imported 2.6 million b/d or 64 per cent of total Third World oil imports.

which is equal to the average per capita income of the entire area of OPEC.[14] These countries' relatively adequate ability to pay their oil bills is such that their development programmes need not really suffer from increases in the price of oil. It is known that their short-term problems of balance of payments and external indebtedness are the result of other factors. More important is that the present high level of energy consumption in those countries reflects a high level of capital accumulation. Consequently, the adverse effects of the increasing oil bill on their balance of payments is counterbalanced by longer-term positive effects on growth, as the bulk of their oil consumption is channelled through the industrial sector which generates dynamic effects on growth and social development. Although with much lesser per capita income, some other energy-intensive countries, like India and Thailand, have relatively balanced external payments and positive terms of trade. Consequently, with a reasonable ability to pay the oil bill, this category of developing country does not in fact suffer from the oil price increases as much as is often suggested. Furthermore, although to a much lesser extent than the developed countries, such countries together with the high per-capita-income developing countries, can benefit from the higher oil prices through the partial recycling of the oil money in increased exports of their goods and services towards the OPEC countries. OPEC's trade with the other relatively highly industrialized developing countries, has substantially increased since the oil price adjustments of 1973/74 (though again to a much lesser extent than with developed countries), thus adding a positive effect on their balance of payments that will mitigate the adverse effects of oil prices.

If these two categories of developing countries are put

14. The average per capita income of OPEC countries for 1977 was $1,033 with a very wide range among the member countries (from a low of $300 and $420 for Indonesia and Nigeria to a high of $14,420 and $12,700 for the United Arab Emirates and Kuwait respectively). This average is much less than the per capita income of some developing countries like Singapore and Argentina whose per capita income is as high as that of Venezuela ($2,820), almost twice that of Iraq ($1,530), more than nine times that of Indonesia and about seven times that of Nigeria.

aside, less than 20 per cent of the total Third World oil bill is borne by the really poor of the poor nations. Because of the very low level of development in those latter countries, energy consumption, and hence oil imports, are extremely low. Only 10 per cent of the total oil imports into the Third World are taken by 54 countries, into each of which the volume of net imports is less than 20,000 b/d. Even with this meagre level of oil imports, oil price increases can create serious adverse effects on the development of those countries, which suffer greatly from balance of payments deficits, external indebtedness and adverse terms of trade. The fact that they do not enjoy the same level of development prevents them from benefiting through increased trade with OPEC, as their ability to recycle the oil money is much less than that of the other developing countries.

But these very same countries are the main beneficiaries of the OPEC aid disbursement. More than two-thirds of this aid has been channelled during recent years to the 45 most seriously affected of the poor countries.[15] For this reason, the least developed countries now receive bilaterally and multilaterally in OPEC aid as a whole more than the increase in their oil bill. Aid disbursements, whether for balance of payments support purposes or development financing, are effected almost entirely on a concessionary basis.

OPEC oil price actions cannot therefore, be held responsible for the development problems of the Third World countries, which suffer more from the existing international economic order and the system of world trade and finance than from oil price increases. The oil bill does not constitute on average more than one-fifth of the total import bill of the developing countries. The other four-fifths are imports mainly from the developed countries, with dramatically increasing costs due to the galloping world inflation which is the main reason behind the persistent adverse terms of trade of the Third World nations. In 1975 their terms of trade had, for example, deteriorated by more than 13 points, at a time when OPEC prices were kept frozen for the most part of that year and then increased

15. This ratio of aid distribution is higher than that of the OECD (*op. cit*) countries.

in October by only 10 per cent. Similarly, in spite of OPEC's frozen prices for the whole year of 1978, those countries' terms of trade deteriorated further by five points.[16] These facts indicate the adverse impact of world inflation —originating in factors other than increased oil prices— on development of the developing countries. Long before OPEC's price actions, developing countries continually suffered in their trade with the developed countries, not only because of the persistent deterioration of their terms of trade and the price of exchanges, but also on account of the protectionist practices of the developed countries against their exports. Tariff and non-tariff barriers, including quantitative controls, government subsidies for similar uneconomic domestic production, etc., have been playing a tremendous role in impeding the growth of export sectors of the developing countries, whether related to primary commodities or to manufactured and semi-manufactured products. UNCTAD's relentless efforts since early 1960 for trade liberalization in favour of the developing countries' exports were all non-productive. Furthermore, the growth of those countries has been impeded by the monopolistic practices of technological transfer from the developed countries and the exorbitant costs of new technology that would help developing countries increase the competitiveness of their manufactured products in world markets.

At the same time, developing countries are suffering from the present international monetary system, in which their increasing indebtedness is dangerously affecting their balance of payments position. Servicing of external debts is beginning to consume a major part of their foreign exchange earnings. Developing countries have been resorting increasingly to the international capital markets and the banking system for borrowing, with ever increasing costs and harsher conditions. In a way, this rising ratio of financial flows from the private sector is the result of the diminishing public aid disbursements from the developed countries who, it is felt by many, saw in the increased OPEC aid

16. Figures about the terms of trade of developing countries are taken from the IMF.

to developing countries a reason to reduce their own aid to the Third World.

International monetary institutions, like the International Monetary Fund, which is heavily weighted in favour of the developed countries in its policy-making process, are further examples of the imbalanced international economic and financial relations of which the Third World countries are the real victims. In addition to the meagre share those countries have of the IMF's lending capacity, the conditions attached to loans, and the severe norms applied in lending money to them, as well as the various other pressures they are subjected to, make them the smallest beneficiaries of such international monetary institutions. The tremendous efforts over the last five years by the Third World nation groups to introduce reforms in the monetary system to improve their position, have all failed because of the intransigent attitude of the industrialized countries.

However, in spite of the imbalance of the international aid performance for development, in which OPEC countries are effectively bearing a disproportionately high burden, the oil producing countries are nevertheless being called upon to enhance further their solidarity with the other countries of the Third World. This requires a clearer and more effective basis for a longer-term strategy, defining OPEC and non-OPEC inter-Third World relationships.

Trade more than aid: Longer-term avenues of economic and commercial cooperation with other developing countries can be found through wider and more permanent channels of expanded trade that go beyond the transitional aid relationships. Temporary by nature as they are OPEC's financial surpluses should not be the only pivot for expanding cooperation with other developing countries. Their surpluses are bound to disappear, as their absorptive capacity of spending on development grows. Hence the impelling need to undertake such structural changes in trade relationships between OPEC and the other developing countries as would help create new opportunities for greater exchanges in the oil and non-oil sectors.

The strategic aim of fostering inter-Third World solidarity as a *sine qua non* for making the international economic order more just necessitates the creation of favourable conditions for higher levels of recycling oil money into the developing countries by increasing the amount of imported goods and services from the other developing countries into the OPEC countries. It should be admitted, however, that this process of recycling is structurally limited by the nature of the existing inter-Third World trade relationship, in which each of the developing countries is more dependent on the developed countries than on the other developing countries. With growing requirements for development, the OPEC countries are also expected to continue to depend heavily on developmental inputs from the developed countries. Nevertheless, part of OPEC's incremental trade could be increasingly oriented towards the imports of goods and services from the other developing countries.

Among the measures that would help change trade structures in favour of more trade with the other developing countries is the investment of part of OPEC's surpluses into the export sectors of the developing countries (primary sectors, i.e. raw materials and agricultural products and food, as well as secondary sectors, i.e. manufactured and semi-manufactured goods) under such conditions as would channel the expanded production of those sectors to the OPEC countries. Favourable conditions should be created in the developing countries themselves to provide the incentives for OPEC countries for such investments.

Effective and selective aid: OPEC's aid effectiveness does not seem to be commensurate with its extensiveness. In spite of its huge aid disbursements in favour of developing countries, OPEC is depicted by many thinkers and policy-makers, even in some of those countries, as failing to meet its political commitments vis-à-vis the Third World. Little is in fact said or known, even among the OPEC countries themselves, about the extensive bilateral aid relationship into which each member country enters individually with various developing countries. So far there is no effective aid policy coordination among the OPEC donors based on common long term objectives. It is time that more clearly defined objectives for aid be laid down and priorities be set for aid disbursements.

The nature of aid differs in accordance with the level of development, balance of payments, terms of trade, etc., of each of the recipient countries. The industrialized developing countries with high national per capita income are not in real need of OPEC concessional aid, as their ability to bear the cost of oil imports is visibly better than that of the other developing countries. However, their short-term external payments problems would warrant some form of cooperation, whereby financial flows from OPEC to those countries could be made on a non-concessionary basis to help them overcome their transitional problems. Some arrangements could be made to encourage those flows by the creation of appropriate financial institutions. It is only the least developed and most seriously affected poorer countries with limited energy imports that are entitled to the concessionary aid, with a greater grants element than other ordinary forms of concessionary loans, such as low interest-bearing loans. Some intermediate solution could also be found for the other category of the developing oil importing countries.

Similarly, the purpose of aid differs from one category of developing country to another, depending on the prevailing development conditions. For the poorer countries, aid to meet their balance of payments deficits would be more justified than in the case of such other countries, where aid could be channelled into long-term development projects.

The institutional aspects of OPEC's aid also need clear definition. It is true that bilateral aid will have to continue to account for the major part of total OPEC aid, as this kind of aid reflects the individual member country's foreign policies and the nature of economic and political relationships with each of the recipient countries. Notwithstanding, there is a need to back-up multilateral aid with common policy objectives. An important part of OPEC aid disbursed through international institutions was less effective in serving OPEC policy than the aid given through multilateral institutions, either those exclusively created by OPEC, such as the OPEC Special Fund, or other multilateral organizations created by certain members of OPEC, such as the Arab Funds, the Arab Bank for Economic Development in Africa, the Islamic Bank for Development, etc. The OPEC Special Fund, which is a collective aid facility that was originally meant to be an *ad hoc* channel to disburse $800 million, was created in 1976. Further replenishments of the Fund's resources, that have brought them to $2.4 billion, together with its successful performance as an effective instrument of collective aid, would strongly justify its conversion into a permanent international aid agency with larger financial resources.[17]

Lastly, OPEC's aid to developing countries cannot really be effective as a contribution to solving problems of international development without being part of a global programme of aid. The increasing aid levels of OPEC should never be a reason to relieve the industrialized countries of their historical responsibility for assisting the Third World countries. On the contrary, it should be used as a means to pressure those countries to fulfil their political commitment for international development. In the final analysis, present problems of development are the product of a world system of trade and finance that was tailored to serve the interests of the industrialized countries at the expense of the developing countries. The old process of resource transfer from the poor to the rich through the price mechanism and trade patterns has been aggravated recently by the world inflation, which became an even more effective instrument of such transfers. Developed countries, therefore, are called upon to increase their aid for development, commensurate with the increase in the price of their exports of goods and services to the developing countries. Iraq's proposal to OPEC for the creation of an international fund for aid, to be jointly financed by the industrialized countries and those of OPEC reflects a new and more penetrating vision towards changing the international

17. At Caracas in December 1979 the oil ministers recommended a further replenishment of $1.6 billion bringing the total financial resources of the fund to $4 billion.

economic order. According to the proposal, the industrialized countries participate in the fund's resources in proportion to the increase of their export prices (world inflation) to the developing countries, whereas those of OPEC would contribute by amounts that are to be determined in the light of price increases of their oil exports to the Third World countries. If it succeeds, this project would bring a real change in the distribution of wealth among nations.*

Cooperation in Energy: OPEC's concern to foster solidarity with the other developing countries is also manifested in the area of energy. The oil producing countries, are according priority treatment to the developing countries in matters related to the security of oil supplies. Measures are currently being taken to ensure that developing countries will be fully provided with their requirements at a cost which reflects the official OPEC price. This aspect of cooperation is extremely important in times of market shortages and price flare-ups in the spot market. In the long-run, securing oil supplies to the developing countries is even more important since the concern over the problem of oil supply to balance world energy requirements will be greater than that over the price of oil. In the long-run also, OPEC's aid for development could be partly linked to the development of domestic energy resources. In such cases, where the potential in the recipient developing countries for promoting those resources exists, part of OPEC aid could be channelled into energy projects to help those countries achieve a higher degree of energy self-sufficiency.[18] In this context, however, it should be borne in mind that assisting Third World nations to develop energy resources, whether conventional or new, is a huge task which requires tremendous financial and technical efforts that could only be achieved by international cooperation in which the industrialized countries shoulder a proportionate responsibility.

* At that Caracas meeting another project for aid was jointly presented by Venezuela and Algeria and was referred by the Conference to the Long-term Strategy Committee.

18. About one-third of the OPEC Special Fund aid was disbursed for the promotion of domestic energy resources in recipient developing countries.

OPEC's Relationship with the Industrialized Countries

Until recently oil prices were at the centre of the economic and political preoccupations of the industrialized countries, and hence a major determinant of their relationship with the oil producing countries, as well as more generally with raw material exporting countries. OPEC price increases were taken as a real threat to their welfare, and as the main cause of their present economic and social problems. Thus OPEC was held responsible for the galloping rates of world inflation, the slow-down in economic growth, the increasing rates of unemployment, the disruption of the world monetary system, etc. Consequently, confrontational and aggressive attitudes towards the oil producing countries built up, and culminated in the creation of the International Energy Agency (IEA) that had the explicit political intention of undermining OPEC.

It is true that OPEC's action on the oil price represents a real departure from the old patterns of prices and international trade relations, in which the industrialized countries were the main beneficiaries. In this sense, therefore, changing the system of pricing oil so radically was a disturbing factor to an economic equilibrium based on the unequal distribution of the benefits of world trade, the correction of which is bound to create difficulties of adjustment in the short-term. The problem, however, was that the oil revolution happened to take place at a time when the developed market economies were undergoing major structural changes. Growth-limiting factors, such as internal inflation, external deficits, rising costs of labour, increasing social pressures, etc., became inherent features of the Western economies long before the increase in oil prices. The change in the nature of the trade and business cycles in those countries was strongly felt during the first half of the 1970s when, unlike the old pattern of fluctuations, galloping rates of internal inflation were associated with economic recession, giving rise to the new phenomenon of 'stagflation'. Likewise, the shake-up of the international monetary system and the collapse of the Bretton Woods arrangements, leading to chaos in the world money markets (manifested among other things by the floating exchange rates and the successive devaluations of the U.S. dollar), were in full swing since the very beginning of the 1970s, i.e. before the oil price increases.

Nevertheless many found in this historical coincidence between the 'oil revolution' and the structural economic changes of the developed countries an easy political instrument for attributing to OPEC the responsibility for the current problems of the world economy. They feared that oil prices would become a new means of draining wealth from the developed countries and would adversely affect their economic growth and capital accumulation to the benefit of the oil producing countries.

However, after it became evident that the industrialized countries were able to live with the higher oil price, and that the effects of the 'oil shock' were absorbed by their economies, more recognition was given to the fact that those structural economic difficulties were not the result of the oil price increases as much as that of internal factors. This has been shown by many studies of the Western countries, which have attributed to the oil price a much smaller role in the economic developments of the 1970s.[19]

On the contrary, we see a new trend in the industrialized countries of favouring an increase in the real price of oil as an effective means of enhancing their efforts towards reducing dependence on imported oil and achieving a reasonable energy mix. It is now realized that only through higher prices in real terms can those countries conserve more energy, curb the growth rates of oil consumption and provide the right incentive to invest in alternative domestic sources of energy.

More recently, however, the real energy concern of the industrialized countries has shifted from the price of oil to that of the security of oil supplies. Following the Iranian oil crisis and the subsequent disruptions in the oil markets, the

19. See for example: *International Trade 1978/79*, General Agreement on Tariffs and Trade, Geneva 1979; *The World Economy in Crisis*, The Securities Group-International Monetary Advisory Board, IMF, August 26–27, 1979; International Implications of OPEC Price Increases, Dr H. A. Merklein, *World Oil*, May 1979.

world awakened to the hard fact that OPEC oil production capacity was not unlimited and able to meet indefinitely the growth of world demand for its oil. Securing future oil supplies from OPEC appeared, therefore, to have far greater importance for sustaining the economic growth of the industrialized countries than the price of oil. Consequently, this issue began to influence more effectively the shaping of their relationship with the oil producing countries. However, it should be borne in mind that for those countries the transitional nature of the problem of security of supply should be considered in the light of their firm policy objectives towards reducing the share of OPEC oil in their total energy balances, the implementation of which would be facilitated by the rising trend of real oil prices.

The concern of the oil producing countries on the other hand is of a longer term nature. Their economic and social development and the structural diversification of their economies constitute the centre of their long-term global oil strategies. For them oil constitutes virtually the only effective means to fulfil their strategic developmental targets, as oil not only provides the financial sources for development programmes and industrialization schemes, but also constitutes the wider base for their industrialization, especially in the area of export-oriented downstream activities, i.e. the hydrocarbon-based investments in the oil refineries, chemical and petrochemical plants, fertilizers, etc. As their internal markets for such products are too small to justify viable investments that could benefit from economies of scale, these countries will have to be increasingly involved in constructing large scale capacities, the production of which will be destined for world markets, in order to secure economic viability for these investments and reasonable returns on capital invested.

The problem, however, is that the technological gap and the commercial policy barriers that separate the oil producing countries from the industrialized countries are among the major obstacles in achieving the objective of industrialization and social change. The industrialized countries monopolize the advanced technologies that are needed by the oil producing countries in their export-oriented investments, whose products require a high level of competitiveness in world markets. The dependence of the oil producing countries on industrialized countries for the acquisition of the necessary technology is virtually total. In many cases the transfer of advanced technology is associated with prohibitive conditions that render it either unfeasible in practice, or so costly that investments become extremely doubtful, if not totally unviable. Furthermore, investments in the export-oriented sectors, (oil and non-oil) are impeded by the commercial practices of the industrialized countries, which prevent market access for processed products from developing countries in general and OPEC countries in particular. It is by tariff and non-tariff barriers, including market-sharing practices, quota systems, quantitative restrictions, etc., that the industrialized countries protect their own production, and hence bar the way to the entry of similar products from suppliers in the developing countries.

The oil producing countries find in such gaps the main causes for the imbalance existing in the pattern of their export-mix. While they account for more than 80 per cent of total world oil trade, only 5 per cent of their oil export took the form of refined products, the remainder being crude oil.[20] Obviously, they will seek every means to correct this imbalance and to try speedily to become exporters of processed products instead of remaining for ever only exporters of raw materials. Such a change in the export-mix is for them a condition for industrial progress and would make their economies benefit from the value added to their crude, capital accumulation, technological change, etc.

On the other hand, OPEC countries are no less concerned about the conservation of their oil resources for future generations. Being developing countries, with mono-structures of national economies and almost total dependence on a single depletable commodity, the oil producing countries will ultimately have to optimize the production and pricing policies of that commodity with a view

20. Compared with a share of two-thirds in the world total production of crude oil in 1977, OPEC's share in the world oil refining capacity was less than 7 per cent and was only 3 per cent of the total petrochemical capacity.

to accelerating their rate of economic development and social change. The lifespan of their oil reserves has therefore to be prolonged to match the time horizon required for achieving their economic and social structural changes. This could be made possible through the adoption of oil production policies, that cope with their development requirements, i.e. to produce such volumes of oil that, at a certain price level, are necessary to generate the required financial resources for development. At the same time, the expansion of the lifespan of those reserves can also be made by enhancing exploratory efforts to find new oil reserves or to continually improve the efficiency of recovering oil from the existing reservoirs (thus adding new reserves) to compensate for the depleted ones.

The fast rates of the depletion of oil reserves in OPEC countries would, if not counterbalanced by new reserves additions at reasonable cost, reduce the incentive to increase production rates in the light of growing world demand for OPEC oil. This factor is of special importance because in the majority of OPEC countries current production profiles are reaching plateau levels and cannot, therefore, be sustained for a long period.

As much as in the downstream, the oil producing countries also need the acquisition of advanced technology in the upstream phase of the oil industry, where they will have eventually to undertake huge investments to add new reserves to compensate for the depleted oil. This is especially so in the case of the complicated techniques of secondary and tertiary recovery; and also in exploration and drilling in the offshore zones.

Furthermore, in meeting world demand for their oil, certain OPEC countries are effectively depleting their resources beyond their real requirements for development. This is specifically the case of the OPEC producers in the Arabian peninsula which, because of their moral obligation vis-à-vis the world economy, are producing more than their needs against paper money, which is depreciating in value over time. The replacement of this depleted oil means high investment risks which, in most cases, are not justified by development priorities. More importantly, this depletion is bound to affect member countries' long-term requirements for oil, both as energy and industrial input for hydrocarbon-based industrialization.

Moreover, for these countries, accumulating financial surpluses in lieu of their extra oil production would mean further risks for development because of the ever increasing vulnerability of these surpluses to the vagaries of world inflation and the chaos in the world monetary markets. It is obvious that without sufficient security to protect these surpluses from various erosion factors on their purchasing power, those countries will definitely seek alternative production policies that will safeguard their interests. No less important a factor in this respect is security against the political risks which those surpluses may face.

Therefore, for the oil producing countries, providing secure oil supplies to the industrialized countries should be counterbalanced with net gains in terms of economic development and oil conservation. In the same manner, just as the industrialized countries need the OPEC countries for oil supplies, those latter countries need the industrialized countries in order to secure the right means for the development of their economies.

In the oil, as much as in the non-oil sector, the interdependence between the oil producing countries and developed countries can be enhanced only when development opportunities are secured for the diversified economic structures through which OPEC countries could reduce their total dependence on oil.

Towards a new long-term OPEC strategy

In their Solemn Declaration made in Algiers in March 1975, the sovereigns and heads of State of the OPEC countries laid down the basic guidelines for an OPEC position concerning the major aspects of a new world order for energy and development. Whether related to the price of oil and the principles for its future evolution or to the other aspects of the energy and development policies and relationships with the consuming countries (developed or developing) the Declaration reflected the lucid and positive thinking of the OPEC countries for an overall solution to the problems of the world economy. It should, however, be recalled that when the Algiers Summit Conference was held, the world 'oil scene' was still influenced by the 'shock' of the OPEC price revolution. Therefore, being mainly concerned with the problem of the oil price movement and its impact on their economic growth, the industrialized countries saw in an international dialogue a favourable means of achieving their long term objectives. On their part the oil producing countries introduced in their Solemn Declaration a global 'package' for a new international economic order that would hopefully be negotiated with the industrialized countries. The failure of the Paris North-South dialogue was the natural consequence of the intransigence of those latter countries which, contrary to OPEC's position, tried to isolate the problem of energy from the other problems of raw materials and development.

The world 'oil scene' had substantially changed since the historic event of Algiers. The world's major energy concern shifted from the price of oil to the energy balances and the role of OPEC crude oil availabilities in the face of rising world demand. The problem of the security of OPEC oil supply as an essential pre-requisite for the sustained growth of the world economy seemed to be at the heart of the economic and political preoccupations of the developed countries. Other aspects of energy and development of the Third World countries, of which OPEC is an integral part, and the clear definition of OPEC's position in relation to them would require formulating a long-term strategy in the light of new conditions. Hence the timely creation in 1978 of the Long Term Strategy Committee formed at the ministerial level by the five founder countries of OPEC and Algeria. The Committee has assigned another technical committee with the same representation and under the same chairmanship of His Excellency Sheik Ahmed Zaki Yamani, Minister of Petroleum and Mineral Resources of the Kingdom of Saudi Arabia, to work on a report dealing with the main issues. After more than a year of intensive work the latter committee finalized its report and, when this is adopted by the ministerial committee, a special OPEC conference will be held to examine it and to come forward with declared principles for a new OPEC long-term strategy.

The Committee's deliberations were centred on the definition of OPEC long-term policies within the new world energy outlook, especially those concerning future supply and demand balances. Among other things, the committee discussed a long-term strategy for the price of oil and the principles governing its future administration by OPEC in the light of the likely world balances. It also discussed the relationship with the other developing countries and the redefinition of OPEC aid policies to the other Third World countries within the larger context of intra- Third World economic cooperation. Finally the Committee studied the relationship with the industrialized countries in the light of new conditions which emerged after the Paris talks.

The foundation of this Committee marks a new, major turning point in OPEC's history that could help enormously in laying down the basic guidelines for future policies covering its main areas of activity. It could also help in the adoption of a common position in any forthcoming international negotiations for a new and just world economic order. Within OPEC as well as outside, great hopes are held for the outcome of the OPEC Long-Term Strategy Committee's deliberations.